Black Radio ... Winner Takes All

Norma & Chris,
Enjoy The History
Marsha Washington George
('RADIO LADY')

Black Radio ...
Winner Takes All

Marsha Washington George

Copyright © 2001 by Marsha Washington George.

Library of Congress Number: 2001117893
ISBN #:   Hardcover   1-4010-2254-5
           Softcover   1-4010-2255-3

All rights reserved. No part of this book may be reproduced or transmitted in any form or by any means, electronic or mechanical, including photocopying, recording, or by any information storage and retrieval system, without permission in writing from the copyright owner.

This book was printed in the United States of America.

Visit the author at www.daradiolady.com

To order additional copies of this book, contact:
Xlibris Corporation
1-888-7-XLIBRIS
www.Xlibris.com
Orders@Xlibris.com

Contents

PREFACE 11

A BRIEF HISTORY OF THE FIRST CENTURY
OF RADIO 13

CHAPTER ONE
   JACK COOPER 37

CHAPTER TWO
   CARLTON MOSS 42
   VERE JOHNS 43
   JOE BOSTIC 44
   GORDON HEATH 46
   ROY WOOD 47
   RICHARD STAMZ 49

CHAPTER THREE
   WILLIAM BLEVINE 55
   HAL JACKSON 57
   CLIFFORD BURDETTE 61

CHAPTER FOUR
   NAT WILLIAMS 63
   MAURICE HULBERT
   HOT ROD 65
   PAUL E.X. BROWN 69

CHAPTER FIVE
   CASANOVA JONES 73
   KEN KNIGHT 77

    THE FIRST BLACK OWNED RADIO STATION 81
    WILLIS SCRUGGS 84
    JOCKEY JACK aka
    "JACK THE RAPPER" 86
    SHELLY STEWART 95
    E. RODNEY JONES 100

CHAPTER SIX
    AL BENSON 106
    FRANKIE HALFACRE 107
    JOHN RICHBOUGH 109
    ALLEY PAT 110

CHAPTER SEVEN
    DONNY BROOKS 113
    TOMMY LEE SMALLS 115
    ZILLA MAYS 118
    EDDIE CASSLEBURY 119
    HERB KENT 134
    BILL "DOC LEE" 135

CHAPTER EIGHT
    JAY BUTLER 138
    SID MC COY & YVONNE DANIELS 141
    YVONNE DANIELS 142
    BOB SUMMERRISE 143

CHAPTER NINE
    EDDIE O'JAY 145

CHAPTER TEN
    LUCKY CORDELL 151

CHAPTER ELEVEN
    JOE WALKER "THE BLIND SCOTSMAN" 157
    HERB LANCE 161

CHAPTER TWELVE
BILL WILLIAMS 164
LEON LEWIS 169
LOUISE WILLIAMS BISHOP 172
JOHNNY ALLEN 172

CHAPTER THIRTEEN
AL JEFFERSON
BUTTERBALL 174
BUTTERBALL (AT THE WAMMY IN MIAMI)
WAME IN MIAMI 175
DOUG STEELE 175

CHAPTER FOURTEEN
IRENE WARE 178
LES ANDERSON 179
BERNARD HAYES 180
RUFUS P. TURNER 187
PAYOLA 199
RADIO RACE RELATIONS INFORMATION CENTER 203
AWAY FROM THE BLUES 206
MITCH FAULKNER 207

CONCLUSION
MY VOICE BECAME MY CRY 216
PIONEER DISC JOCKEY
HONOR ROLL 217
FAIR PLAY COMMITTEE FACT SHEET 219
DEE JAY TRIVIA 222
WORKS CONSULTED 226
INTERVIEW CREDITS 229

INDEX 231

Dedication

IN MEMORY OF KEN KNIGHT - MY BELOVED UNCLE WHO'S RADIO CAREER GAVE ME THE REASON TO WRITE THIS BOOK. TO THE MEMORY OF RCA RECORD PIONEER GENE BURLESON WHO GOT ME THROUGH THE TOUGH TIMES. TO THE MANY RADIO PIONEERS WHO WENT THROUGH CLOSETS AND FILES TO GIVE A GRAND STORY. AND, TO MY FAMILY FOR YOUR ENCOURAGEMENT.

PREFACE

The significance of being the first of anything is an achievement in itself. In all of the masterful references of who's the first in the various sectors of American History, not one part of literature has given a true salute to those who opened gateways for our communities. Never has anyone decided to look at the scope of a generation of brave ones such as these. The ones who had no one to turn to, only their very own imaginations when the need for responsibility called.

Although many of these brave pioneers are no longer with us to enjoy the salute to their challenges and successes, they do however remain a legacy in history.

I hope that the memories given in the following pages will give them a place in our hearts and in our past, present, and future history.

I shall now begin this journey through the paths of their lives. I invite you to share the history as it is truthfully told for the first time.

Together we will examine their charismatic personalities, and we will all see that this story is also ours.

A legend in itself is this journey. A journey of business, compassion, dedication, friendship; their journey—our leaders of broadcasting. I invite all to read and share this material of a rich history whose time has come. I invite you to share with the Disc Jockeys what it meant for you to hear their voices which echoed across our neighborhoods, and their music which became our mourning sun.

Imagine, if you will a little girl sitting comfortably on the floor while looking up at the radio speaker with her younger sister, only to hear the voice of her beloved Uncle speak . . .

Thus begins her telling of: America's 1st Black Disc Jockeys, a historical journey from the beginning.

A Brief History of the First Century of Radio

During the embryonic stages of radio there was quite a thrill of wonderment experienced by people of the 1920's when they listened to the prolific voices for the first time. The embellished mixture of voices and music coming over the radio seemed almost supernatural. The excitement was compounded when other distant stations could be heard.

It was in these earlier years of broadcasting when the programs didn't matter so much. Initially, it was satisfaction enough to have these far away voices permeating the atmosphere of your own home.

This amazing age of musical discovery also brought fourth another industry credit when a noted industry icon, Dr. Conrad initiated the term record promotion. Record promotion began when Hamilton Music Store offered to keep Dr. Conrad supplied with new records. Upon his acceptance he agreed to announce that the records came from Hamilton Music Store. Thus, making the first commercial advertising deal in the industry which became known as a "trade out", a practice still in use today.

Later the Joseph Horne Co. of Pittsburg started to advertise the sale of radio receiving sets, priced $10 and up, so that people could hear Dr. Conrad's Victorola "concerts". Westinghouse Vice President H.P. Davis saw many profit possibilities in this new marvel and used the experimental station 8XK which became KDKA and went on the air November 2, 1920, with the returns of the Harding-Cox Presidential election.

The Detroit News had been operating a radio telephone for several years and was licensed in 1921 as WWJ. In addition, KQW in San Jose (now KCBS) started on the air in 1909 and begin a regular schedule in 1912.

The first sports broadcast of consequence was the Dempsey-Carpentier fight on July 2, 1921. The first live orchestra program was in 1921, when Vincent Lopes bought his band to the WIZ studios in Newark as a favor to his friend Tommy Cowan, the program director. The musicians unfortunately were not paid, but, they received the opportunity to promote themselves. On that same program was a young baritone, John Charles Thomas, seeking to publicize his appearance at a local club.

In 1922 Tommy Cowan hired a young tenor as a WIZ announcer. His name was Milton Cross, later to become the official voice of symphony and opera. Network radio was born in 1926, when the National Broadcasting Company was organized by Radio Corporation of America. NBC later split into the Red and the Blue networks, WEAF and WJZ their key station in New York. Young William S. Paley, in 1929, merged the Columbia Phonograph Broadcasting Company with the United Independent Broadcasters and formed the Columbia Broadcasting System, with WABC in New York (now WCBS) and 46 affiliates.

The first coast to coast network broadcast was the 1929 Rose Bowl game, described over NBC by Graham McNamee.

During the 30's, New York and San Francisco were the key origination points for network shows, with Chicago also supplying a good many. The NBC studio in Los Angeles was a small building on the RKO movie lot. CBS originated a few network shows from the KHJ studios in the Don Lee Cadillac building. Most Los Angeles organizations in the 30's were similar to those of the big name shows, normally done in New York, with stars, such as, Rudy Valee, Ben Bernie, etc. who began to become a part of the movie screen . In the early 1940's NBC built it's own studios at the corner of Sunset and Vine, and CBS put its studios a few blocks farther up Sunset. Shortly, thereafter, Hollywood took over the West Coast origination of the big network shows, while NBC's new San Francisco studios, four floors tall, were increasingly devoted to sustainers and to local programs. Subsequently the Blue Network separated completely from NBC and became the American Broadcasting Company.

In the 30's came the great depression but radio was hardly aware of it. Big console sets dominated millions of living rooms, just as TV sets do today. Sponsors were frequently amazed at the effectiveness of their commercials. There was little need for audience measurement surveys in the early days: a program's popularity could be rated according to its mail count, regularly tallied by the fan mail department. Your mail count was as important then as Pulse, or ARB numbers are today.

However, it wasn't all network radio. WNEW had Martin Block and KFWB had Al Jarvis playing popular records, and their fan mail often rivaled that of the network shows.

When World War II began it was reported by many radio correspondents. The strong voice of Edward R. Murrow reporting bombings from London's rooftops could be heard over the radio. You could also listen to messages from President Franklin Roosevelt as he spoke to everyone from his fireside chats. This method of communications became a part of family life in a way that television can never be compare to.

In the early 1950's television began to come into play and take away many of the national advertisers. Thus causing a panic among some of those in radio who also believed that radio would only be filled with news, weather and time signals as a result of this new found technology known as TV. But they were wrong as this newly developed piece of technology was enveloped by a new breed, the disc jockey. In came men like Bill Randle, Buddy Deane, Ed McKenzie, Al Benson, Hal Jackson, Tommy Smalls, Alan Freed, Ken Knight, Jack Gibson, and others as they began to make hit records and thus caused a new creation of radio phenomenon.

And as communication is improved with each newly developed technological discovery, the disc jockey began to have control of which format could really make things happen for their station and their career. And with this comes looking at where the disc jockey has been in the past and what's ahead for them in the future.

An article written by Sam Chase in his Music Business Magazine. highlights their stamina as "a willingness to run with the record and take a chance on a new artist" as he recognized the black disc jockey for having made such a powerful influence on the pop music scene. It was during the 1950's when rhythm and blues made it's first real impact on the pop market. From a specialized music appealing to only a small segment of a population, rhythm and blues spilled over into the pop market and fulfilled a pin up demand for music with a beat that has never subsided to this day. One of the early rhythm and blues records that turned into a pop hit in the 1950's was Sh-Boom on the Atlantic label, this recording sold close to a million copies made the top ten on most of the trade paper charts. Another was Gee on the RAMA label, where many of the rhythm and blues that had gone pop in the early decade, but, these had occurred only sporadically. The 1950's R&B deejays move into the pop market turned into a veritable flood. Dave Dreyer one of the pioneers publishers of rhythm and blues song who published, `Sugar Lump' and `Such a Night' during the 1950's had an impact comparable to that of the swing bands in the 1930's. As R&B records broke into the pop market at first in 1952, 53 and 54 many of the A&R men at the larger labels would cover the original hits with strong artists and opt to come through with the best selling record; but, as the record buyer became more familiar with the top R&B names, the cover record passed by the way side and original discing took all the honors, sales, place and attention. Then, as now, the R&B disc jockeys and the R&B stations provided the initial exposure for these recordings. As the R&B wave built to a taunt many of the pop stations took their lead to the hot local R&B stations as to what to put on their play lists. Rhythm and Blues became Rock n Roll with the hottest of the disc jockeys at that time in 1955 with Alan Freed at WINS, who started to call the records he played Rock `n' Roll. By 1960 Rock and Roll was so firmly established that even the die hards in the music business finally realized that it was here to stay. Through the 1950's it was the R&B disc jockeys who started the hits, by the 1960's this became more true than ever when pop stations tightened up their procedures so that new records had to be listened to by a committee before they could be put on the air, record manufacturers never happy about waiting to get their records aired turned more and more to the R&B deejays, who worked on smaller stations with less staff could jump on a record quickly and could get on the air ahead of the top 40 stations. Today, the influence of the R&B jocks is at it's highest point. More and more records hits get their initial exposure and have their way to big sales, promoted by the air play they receive at the hands of the rhythm and blues disc jockey. No longer are they neglected by the large companies and they never were neglected by the small. They have probably started more artists on the road to success than any other group of disc jockeys in the country. All in all the Rhythm and Blues Disc Jockey is a powerful voice on the american pop scene.

Another article written by Sam Chase covers the importance of local surveys for R&B radio station as it was a well integrated part of the famed radio and exposure chart published in music business each week. "We think this sets a significant presidence in the recording industry. The importance of R&B disc jockeys and stations has long been recognized by numerous publications through the use of R&B charts or columns, but, to us this is no longer enough. We believe this segregation of R&B is behind the times historically and inadequate from the standpoint of the record industry. The study of so called R&B charts reveals very few titles that do not appeal on so called pop charts. By the time a record is hot enough to hit these R&B charts they have also been picked up by the formula pop stations and have hit the bottom of the pop charts as well. For this reason, a segregated R&B chart no longer makes much sense. The primary significance of the R&B Jockey and R&B radio station is not that they play a different kind of material. Today, in fact the so called R&B material has made the strongest showing of all American music in the face of the current influx of music from Britain and elsewhere. The real importance of the R&B Jockey and radio station is that they are able to take the lead in breaking new hits because they are not limited to play list or formulas. The inclusion of R&B station list and our radio exposure chart, makes it possible to study the position of records on these list and to pin point new records that have not begun to show up on the standard top forty list. To see exactly what we mean we urge readers interested in what's happening to check this weeks radio exposure charts for the surveys of KATZ, St. Louis, WAMO Pittsburgh, and WRAP Norfolk or last weeks chart of WCIN Cincinnati, as what the other stations use we will rotate outlets each week. To sum up we believe that in the record business as well as in the nation that the future lies with integration, not segregation." says, Sam Chase.

But, before we look into the future of the black disc jockey and their peers, we must study the past of the black disc jockey and record in our minds exactly what it was to be air jockeys which they became known as. Understanding this disc jockey of the past permits us to also understand what styles of music was played by them. Along with studying the style of artist(s) that were a part of their present day.

On the following list one can easily see the types of songs aired during the beginning of their era which almost instantly gave them a status symbol. This specialized review of what Bill Board selected as the number one hits of the day is only a prime example of what types of production selections they were up against in keeping the audience satisfied. This historical listing is a backing of their careers as their talents unfold in the various interviews given.
Rhythm & Blues

In order to understand the capacity of pressure building up in the field or radio we must look back at how rhythm and blues surfaced in the industry and thereby incorporate the formats of change which these disc jockeys had to creatively abide by.

After World War II, only a few radio stations were playing the blues and the ones committed to this market were located mostly in the northeastern cities with only an occasional to be found in southern cities. As the request of rhythm and blues became more and more in demand in 1951, deejays were beginning to play more and more of what their listeners wanted to hear, i.e. rhythm and blues. In the

South things were also beginning to change because some of the stations were owned and operated by Blacks and because of this the Southern region was becoming a much stronger point of operation in the industry.

At one time there had been a division between rhythm and blues vs pop, and Dee Jay's were having difficulty deciding whether they should include pop music in their format. This was all due to the fact that rhythm and blues releases were breaking more and more into the industry and pop wasn't. The Dee Jays were turning over releases again and again and after an artist's second release was made, if it didn't make it you weren't selected anymore, which, was putting pop music out of the running. The Disc Jockeys were starting to use this new trend of music everywhere, from coast to coast. There were so many selections for them to play they were beginning to have problems deciding which records to play.

Not only was the music changing, but, the buying audience changed as well. From an age bracket of 25 in 1952 to an age bracket of 18 in 1955. The youth were beginning to control the air play, and the air play for them was rock 'n' roll, and rhythm and blues.

Although, this new trend was taking place, there were cities like Boston which banned rock 'n' roll, thus, began censorship. One of the larger cities that had to control this problem was Chicago. There were 15,000 letters sent to station WGN by listeners demanding that objectional lyrics, sexual incenuendos, drinking, and so fourth be excluded from the station's air play. Thus putting Chicago in the position of having to establish censorship boards which would ban a song like 'Teach Me Tonight'. In business establishments where juke boxes were located the local police would make sure that the songs censored by the board weren't played, as they placed list of the banned songs on them. Places like Bridgeport, and New Haven, Connecticut banned dances which centered around the rock 'n' roll theme as a result of parents complaining.

This new law of censorship bought more challenges for the disc jockeys. Disc Jockey's like Tommy 'Dr. Jive' Smalls and Hal Jackson started staging week long concerts at places like the Apollo Theater and the Brooklyn Paramount, and on occasions they used the Academy of Music.

Of course this was working out fine for the Disc Jockey's but booking agents weren't too happy because this was taking away their promotional arrangements. However, they couldn't complain too much, remember, the disc jockeys had control of unlimited air time to promote shows which were coming to town.
If disc jockeys didn't get big name talents, then the promoters didn't get air play from their station. (kind of a catch 22? I don't think so, I call it survival).

At this time in 1955, the first Rock 'N Roll hit "Rock Around The Clock" by Bill Haley and the Comets hit number one. And not only was this history being made, but, the history of the south was also occurring when a negro seamstress was arrested in Montgomery for refusing to give up her seat on a metropolitan bus to a white man . . . and the civil rights movement was beginning to start as Martin Luther King, Jr. organized a boycott.

(with all of the southern uproars beginning to take place, one would think that, maybe, Rhythm and Blues may have been more of a soothing solution to comfort souls, rather than, a problem.

As the 1950's included more and more entertainers for the public's eye, so did the inclusion of more and more Disc Jockeys begin to surface. The impact of their connections made things happen. One example would be that of Jack Gibson, (then known as

Jockey-Jack), and Eddie Casslebury when they were giving one of their many shows at the Regal Theater in Cincinnati, Ohio. Ike and Tina Turner were the headlines of the show and Ike's band was to play for the show. Just before the first show time they got a call saying the band had been in an accident outside Columbus. Jack and Eddie went over to King Records where James Brown was recording and asked the 'Godfather of Soul', if he would help stall the crowd at the Regal until Tina and Company got there. Mr. Brown brought his entire band and did his dynamite show for almost three hours and didn't charge them a cent. They got the James Brown and Ike and Tina Turner Revues for $750.00. "Things like this just couldn't happen today" it was reported to have been said by most.

But what was happening was a watchful eye was being kept on the status of positions opening up for minorities entering into the industry as documentation of industry actions was being recorded. In 1970, the Radio/Race Relations Information Center reported their findings . . .

The findings of the Race Relations Information Center of Nashville was a detailed study of black-aimed radio programs entitled, "How Soulful is `Soul' Radio?"

The center is a private non-profit organization that collects and distributes information about race relations in this country, and is headed by Luther Foster, president of the Tuskegee (Ala.) Institute. John Seigenthaler, editor of the Nashville Tennessean, is vice chairman.

"Over-all," the report states, "soul" radio's responsiveness to the black community showed a marked increase in the sixties, with the greater changes occurring in public affairs, advertising, news broadcasting and equal opportunity. But few broadcasters showed any willingness to move until prodded by black sentiments (and new Federal Communications Commission vigilance) and some still have moved only very slowly." The report says it can be expected that pressure on broadcasters from the black community will increase "forcing either a boom in black radio entrepreneurship or radical changes in white broadcaster's policies."

A recent development has been the formation of the only black-owned radio chain in the country, the result of three station purchases by singer James Brown. Mr. Brown purchased WRDW in Augusta, Ga., WJBE in Knoxville, Tenn., and WEBB in Baltimore. Each of these stations is well-rated both in studies of black audiences and in general station surveys.

The report lists Mr. Brown's three stations and only six others as owned by blacks among the 310 stations in the country that program for black audiences. These are KPRS in Kansas City, Mo., KWK in St. Louis; WCHB and WCHD-FM, in Raleigh, N.C. Other estimates of black-owned stations have been higher. Last month, the magazine Advertising Age listed the same stations as those cited by the center, but added WMPP in Chicago; WEUP in Huntsville, Ala.; WTLC in Indianapolis; WOR-TV in Hattisburg, Miss.; WWWS-FM in Saginaw, Mich., and WVOE in Chadburn, N.C.

The center studied the programming and hiring practices of five white-owned black-oriented broadcasting chains and concluded that there appeared to be some instances of blacks being given titles but no responsibilities.

The five chains have 22 stations, including Rollins, Inc. Broadcasting Division, and Rounsaville Radio Stations in Atlanta; the Sonderling Broadcasting Corporation in Nashville; Speidel Broadcasters, Inc., in Columbia, S.C., and the United Broadcasting Company in Washington.

The survey found that hiring practices at Rounsaville and Sonderling were the fairest of the chains, and added: "Even some critics who argued that no broadcaster can take pride in his personal record acknowledged in the achievement of these two firms."

Each of these stations relies heavily on contemporary, popular rock music and rhythm and blues. Each one says it broadcasts this music in response to listeners' demands, but, critics, the report states, dispute this. William Wright, director of Unity House in Washington, says "the problem comes with white broadcasters who have brainwashed black people into accepting 24 hours of `soul.' They've created a `soul music' mentality, and black kids are paying for it."

Nearly a quarter a century after a radio station first geared its entire broadcasting format to black interests, there still isn't a nationwide black-oriented news network," the report says. "Blacks still comprise the vast minority in key executive stations at `soul' stations. . . . All this troubles black radio reformers. In addition, they wonder if white management's public affairs efforts are meant to enlighten and serve the masses or merely to satisfy minimum F.C.C. requirements and remain in business."

And while the study given above was being reviewed radio station WRVR FM outlet of the Riverside Church by the Television, Radio, and the Film Commission of the United Methodist Church along with the National Council of Churches, the National Catholic Office of Radio and Television and the National Urban Coalition began to establish something the community was longing for with a program entitled, 'Night Call', which was designed to allow callers from the Nation's ghettos to call in and speak with their prominent guest speaker. As the community began to learn of this program they were informed that they could call in and speak with the guest speaker during the hours of 11:30 pm to 12:30 am to discuss the various topics. As the station tried to assist those from the communities from having to pay for long distance calls, the telephone company objected because they feared that too many lines would be tied up, as a result, the station developed plans on how to reimburse each caller. The 'Night Call' host was pioneer disc jockey Dale Shields from radio station WLIB in New York. The first guest for the show would be the Rev. Ralph Abernathy, acting Director of the Southern Christian Leadership Conference, and others such as Stokely Carmichael, author of 'Fail Safe', Harvery Wheeler and others.

If other stations would decide to do something of this caliber then there wouldn't be any need to do studies such as the one above.

Some of the Songs Played By Disc Jockeys From 1938 to 1965

Name	Song Title	Weeks at No.1
Sammy Kaye	Rosalie	2
Duke Ellington	I Let a Song go out of my heart	3
Ella Fitzgerald with Chick Webb	A-Tisket, A-Tasket	10
Tommy Dorsey	Music, Maestro, Please	6
Sammy Kaye	Love Walked In	2
Benny Goodman	Don't Be That Way	5
Fred Astaire	Change Partners	2
Bing Crosby and Connee Boswell	Alexander's Ragtime Band	2
Fats Waller	Two Sleepy People	2
Bing Crosby	You Must Have Been A Beautiful Baby	2
Al Donohue	Jeepers Creepers	5
Guy Lombardo	Penny Serenade	1
Tommy Dorsey	Our Love	1

Les Brown	Sentimental Journey	9
Betty Hutton	Doctor, Lawyer, Indian Chief	2
Johnny Mercer	Personality	2
Frankie Carle	Oh! What It Seemed to Be	11
Ink Spots	The Gypsy	13
Ink Spots	To each his own	1
Nat King Cole	For Sentimental Reasons	6
Count Basie	Open The Door Richard!	1
Doris Day	A Guy is a Guy	1
Percy Faith	Delicado	1
Rosemary Clooney	Half as Much	3
Patti Page	I Went To Your Wedding	10
Mills Brothers	The Glow Worm	3
Jimmy Boyd	I saw Mommy Kissing Santa Claus	2
Patti Page	The Doggie in the Window	8
Percy Faith	Song From 'Moulin Rouge'	10
Eddie Fisher	I'm Walking Behind You	7
Perry Como	No Other Love	4
Ames Brothers	You, You, You	8
Tony Bennett	Rags to Riches	8
Jo Stafford	Make Love to me!	7
Chordettes	Mr. Sandman	7
Fontaine Sisters	Hearts of Stone	3
McGuire Sisters	Sincerely	10

Black Radio ... Winner Takes All

Georgia Gibbs	Dance with me Henry	3
Les Baxter	Unchained Melody	2
Sammy Davis,Jr.	Starring Sammy Davis Jr., Album	6
Pat Boone	Ain't That a sham	2
Dinah Shore	Anniversary Song	2
Harmonicats	Peg O' My Heart	8
Tex Williams	Smoke! Smoke! Smoke! That Cigarette	6
Vaughn Monroe	Ballerina	10
Peggy Lee	Mañana (Is Soon enough for me)	9
Nat King Cole	Nature Boy	8
Ken Griffin	You Can't Be True Dear	7
Pee Wee Hunt	Twelfth Street Rag	8
Dinah Shore	Buttons and Bows	10
Doris Day and Buddy Clark	Love Somebody	5
Evelyn Knight	A little Bird Told Me	7
Red Foley	Chattanoogie Shoe Shine Boy	8
Ames Brothers	Sentimental Me	1
Nat King Cole	Mona Lisa	8
Patti Page	All My Love	5
Nat King Cole	Too Young	5
Johnny Ray	Cry	11
Georgia Gibbs	Kiss of Fire	7

Benny Goodman	And the Angels Sing	5
Kay Kyser	Three Little Fishes	2
Wil Glahe	Beer Barrel Polka	4
Glenn Miller	Moon Love	4
Glenn Miller	Over The Rainbow	7
Ink Spots	Address Unknown	1
Frankie Masters	Scatter-Brain	8
Glenn Miller	When you wish upon a star	5
Glenn Miller	Tuxedo Junction	9
Glenn Miller	Imagination	3
Tommy Dorsey	I'll Never Smile Again	12
Ink Spots	We Three (My Echo, My Shadow, and Me)	3
Sammy Kaye	Daddy	8
Mills Brothers	Paper Doll	12
Andrews Sisters	Shoo-Shoo Baby	9
Louis Jordan	G.I. Jive	2
Mills Brothers	You Always Hurt The One You Love	5
Dinah Shore	I'll Walk Alone	4
Ella Fitzgerald and the Ink Spots	Into Each Life Some Rain Must Fall	2
Ella Fitzgerald and the Ink Spots	I'm Making Believe	2
Les Brown	My Dreams are getting Better all the time	7

Percy Faith	Theme From a Summer Place	9
Brenda Lee	I'm Sorry	3
Chubby Checker	The Twist	2
Drifters	Save The Last Dance For Me	3
Ray Charles	Georgia On My Mind	1
Elvis Presley	G.I. Blues	10
Soundtrack	Exodus	14
Shirelles	Will You Love Me Tomorrow	2
Chubby Checker	Pony Time	3
Ernie K-Doe	Mother-in-Law	1
Ricky Nelson	Travelin' Man	2
Bobby Lewis	Tossin' and Turnin'	7
Ray Charles	Hit the road Jack	2
Dion	Runaround Sue	2
Jimmy Dean	Big Bad John	5
Marvelettes	Please Mr. Postman	1
Gene Chandler	Duke of Earl	3
Shirelles	Soldier Boy	3
Ray Charles	I Can't Stop Loving You	5
Ray Charles	Modern Sounds in Country and Western Music	14
Bobby Vinton	Roses are red my love	4
Little Eva	The Loco-Motion	1
Four Seasons	Sherry	5

Artist	Title	
Elvis Presley	Hound Dog	11
Platters	The Great Pretender	2
Harry Belafonte	Belafonte 'The Album'	6
Elvis Presley	Heartbreak Hotel	10
Pat Boone	I Almost Lost My Mind	4
Platters	My Prayer	5
Harry Belafonte	Calypso 'The Album'	31
Elvis Presley	Love Me Tender	5
Guy Mitchell	Singing The Blues	10
Andy Williams	Butterfly	3
Nat King Cole	Love Is The Thing	8
Paul Anka	Diana	1
Crickets	That'll Be The Day	1
Sam Cooke	You Send Me	3
Silhouettes	Get A Job	2
Laurie London	He's Got The Whole World In His Hands	4
Everly Brothers	All I Have to Do is Dream	5
Johnny Mathis	Johnny Greatest Hits	3
Coasters	Yakety Yak	1
Chipmunks	The Chipmunk Song	4
Platters	Smoke Get In Your Eyes	3
Henry Mancini	Music from Peter Gunn	10
Lloyd Price	Stagger Lee	4
Johnny Mathis	Heavenly	5

Petula Clark	Downtown	2
Righteous Brothers	You've Lost That Lovin' Feeling	2
Temptations	My Girl	1
Supremes	Stop In The Name of Love	2
Supremes	Back in My Arms Again	1
Four Tops	I Can't Help Myself Sugar Pie Honey Bunch	2
Sonny & Cher	I Got You Babe	3
Rolling Stones	Out of Our Heads-Album	3
Supremes	I Hear A Symphony	2

Bobby Pickett	Monster Mash	2
Four Seasons	Big Girls Don't Cry	5
Steve Lawrence	Go Away Little Girl	2
Rooftop Singers	Walk Right In	2
Paul Anka	Hey Paula	3
Ruby and the Romantics	Our day will come	1
Jimmy Soul	If you wanna be Happy	2
Essex	Easier Said Than Done	2
Stevie Wonder	Finger Tips - Part 2	3
Stevie Wonder	Little Stevie Wonder The 12 year old genius	1
Angels	My Boyfriend's Back	3
Beatles	I wanna hold your hand	7
Beatles	She loves you	2
Beatles	Can't buy me love	5
Louis Armstrong	Hello Dolly	1
Mary Wells	My Guy	2
Dixie Cups	Chapel of Love	3
Beatles	Hard Days Night - Album	14
Supremes	Where Did Our Love Go	2
Supremes	Baby Love	4
Shangri-Las	Leader of the Pack	1
Bobby Vinton	Mr. Lonely	1
Supremes	Come see About Me	2

PART TWO

THE FIRST BLACK DISC JOCKEY

CHAPTER ONE

JACK COOPER

The Beginning of it all . . .
The First Black Disc Jockey

Unfortunately, none of our history books explained the importance of new technology and challenges put before people of color. These very people decided to use their talents and expose to the world exactly what they could do under very oppressive circumstances.

And by eliminating this very portion of history, we were all never introduced to the person who has been recognized as the first black Disc Jockey, who was known to the Chicago community as Jack Cooper. During the time in which Jack Cooper broadcasted there wasn't any recognition being given to the black community as a whole, but, he managed through his determined abilities to change the settled pattern as explained below.

As Jack Cooper began to introduce Black oriented radio

programming in 1929 at Chicago's WSBC radio station, the community began to awaken with applauds of accomplishments. Because we were finally getting a chance to be heard over the airwaves, and our voices and opinions were finally beginning to count. Almost, immediately, Jack Cooper began to buy and sell time through brokerage arrangements. This method of selling was more than enough for the station to allow him to remain because the community was listening and buying the products that were being advertised at the same time. It was also a surprise to know that his show was also commercially sponsored. And as quickly as he began to receive commercial sponsorship his first shows of fifteen minutes quickly moved to an advanced thirty minute program which featured live performances of local Black choirs, organist, recorded jazz, news, public service announcements, and interviews with outstanding Black personalities. Yes, Jack Cooper's efforts were waking up the neighborhood of Chicago and the city was never the same again.

And with each horizon of accomplishment Jack Cooper began to expand his show to one hour long. Jack Cooper recognized the need of his position for the community and thereby billed his show as the "**Negro Hour**." Four years later he was airing five and a half hours of Black oriented programs. And as he completed this milestone in 1935, the community supported him wholeheartedly by coming to see him do his broadcasts and listened to him on a continuing basis.

It has been said over and over again that Cooper's shows were directed toward whites and middle-class Blacks. But, this was not a negative criterior in 1929. The community was proud of him, and supported him without hesitation. Due partly, to the fact that this was a time when all blacks from all walks of life wanted to be considered middle-class and prove to others that they too, could shine and demonstrate that they were also from the better and comfortable communities. Jack's announcing style may have been modeled after that of his white counterparts, (as some have been quoted to say), but;

we must understand that this was the only example he had to use. Afterall, this was the period of the late 20's and early 30's, and radio was a new and amazing phenomenon for everybody. Jack Cooper and his other satellite announcers used in their broadcast the communication patterns of midstream America. And it may have been a little recognizable to the white side as some may have been quoted as saying. But, understand it worked and opened doors for others to become a part of an amazing industry.

Of course, questions were often asked by the newer announcers that later became a part of the industry to the effect of: Why did they use the speech patterns of main stream America and not their own culture voices? And of course some of us tried to answer, such as, Norman Spaulding; an announcer from Chicago, who tried to give a meaning why? by saying, "the earlier Black Disc Jockeys "were not Southern born, or `they' had left the South at an early age, and had no trace of a Southern accent." Unfortunately, Norman Spaulding didn't have the answer to the question and tried to answer it to the best of his ability. But, it would take looking a little deeper into the situation as we find there were many from the South who didn't give up their formal dialect which they were raised with. It was only another case of the types of stereotype many people put on southerners during this time as they also do today. This very thought of one announcer from the past is given as the only answer to describe the methodology of some of the pioneers of black radio, and, thus gives us another reason as to why we should investigate our history of this time even more. Also, why we should try and discover for ourselves whether or not anybody was from the South who may have tried to hide their ancestry. We shall see if these first pioneers denied themselves of a heritage or culture, or were they trying to be the best at what they were attempting to do in the field of broadcasting.

As they also tried to erase the earlier portrayal placed upon the race with the captions of the Amos N' Andy images. They knew that they had too much responsibility and to get on the air waves at this time they truly didn't want to sound ignorant. But this doesn't mean that they were afraid to talk like the community they came from, be it from the South, or, anywhere else as we shall later see.

Accepting the fact that this was a time of survival, there were many problems besides how powerful someones' broadcasting voice was for the community. Not to mention the fact that some stations wouldn't hire you if you didn't sound white. And as from being from the South, there's no proof that regional parts of the country determined how they would react if an announcer would come to the North to work. The boss made the decisions for you, and some Disc Jockeys lasted and others didn't and that's exactly how it was.

Race was an issue of the day; thereby, establishing a standing ground for racial pride as it became an instant concern of the Black announcers who influenced the community with the type of music they played. As one announcer described it, "the first Black Disc Jockeys played the better type of jazz records, and did not play that low-down gut bucket blues."

Ah! but could this last forever? The mid-1940 Disc Jockeys were of a different breed. There were a new group of Black Disc Jockeys which emerged. They were among southern migrants who settled in the northern industrial cities during World War II. This group revolutionized Black-oriented radio through the introduction of Black idiomatic expressions, jargon, and improvised "street talk" communications, as the lower economic class was finally being recognized, and thereby began the historical plight of programming for blues, gospel, and rhythm and blues. (Something more of us can relate to . . . it surely beats the really gutsy images of Amos N' Andy as previously portrayed by some of the earlier radio broadcasts).

As black-oriented radio began to develop more and more the "rapping" Disc Jockeys began to dominate the waves, thereby, overshadowing the announcing style of the earlier Black announcers.

The music programs and the announcing style of these Post-War Disc Jockeys reflected the musical preferences, cultural values and social practices of southern Blacks, many of whom now resided in urban ghettos of the North, (as some have called their communities during this time).

But, has much changed since their beginning? Has the audience demanded more or less from them since the days of Jack Cooper? Is what we're hearing what we really want to hear? Is it racial, human spirit, dedication or determination?

It's up to you to determine as you examine the history of the times. . . .

As the previous information entertained you with the thoughts of what some people had to say about Jack Cooper and Disc Jockey's from the South. It must be reemphasize because, it is very important that we all take a full study at what Jack Cooper actually was involved with and learn what has come to be known as recorded history for he and his listening audience as they read the daily newspapers covering his broadcast during 1929. Today, we have an imaged view of what Jack Cooper stage looked like and what he did with his audiences when we view Tom Joyner's morning show on stage in the cities he and his entourage visit.

CHAPTER TWO

CARLTON MOSS

In early 1931 Carlton Moss joined the masses in the radio industry with his show entitled, "Careless Love,". His complete music broadcast was heard over station WEAF, beginning at 9:45 on Friday evenings, to one half hour, beginning at 7:30 on Monday evenings.

Carlton Moss, received his education from his alumni Morgan College and used his training to project the talents of such noted actors and performers as Frank Wilson, Lewis Thomas, Clarence Williams, Eva Taylor, and Georgia Burke. The incidental music which was aired was furnished by Mr. Williams, Miss. Taylor, Miss. Burke and a quartette known as "The Southernaires." And this timely show was not only produced by Negroes (as once quoted), but, was about the race, performed by people of color, and was also one of the most amusing half hours on the air as it has been duly recorded. Need I add without the horrific interpretations of Amos `N' Andy type of images.

And as the images were steadily being dissolved into one of more constant courage and stamina the community of Harlem was recognized with a voice of it's own. As Vere Johns became the first Colored Harlem Reporter.

VERE JOHNS

Vere Johns weekly broadcast began as news reports for the New York Age at 4:30 pm on Tuesday's over radio station WGV. Even though he has been recognized as the first colored reporter in Harlem, Mr. Johns, was a native of Mandeville, Jamica, B.W., and was educated in the islands. He was also an active member of the British West Indies regiment in the World War and served in Egypt and other places which finally placed him as a recipient of the honor of decoration for bravery. Vere Johns held positions in the Jamaican government, a position in the post office department and the final position held in Jamica was his being a statistical clerk in the revenue department.

Later, in 1929, he came to the United States and began to engage in journalism work and broadcasting. His appearance on stage in classical roles regarded him as an elocutionist of vocal projection. His broadcasting career spanned throughout the years at stations WJZ, WDCA, and WABC. During this time it was also noted that he lived on 113th street, (addresses of artists were given out doing this time period). While maintaining a position as the radio editor for the Age he also wrote dramatic criticisms for the paper.

As the street name recognition is mentioned above we must be reminded that a lot of changes came about from this time period to present day with the resulting factor being that addresses or streets which Disc Jockeys live on could no longer be listed. Today's status of the disc jockey won't allow it, and today's society is much more different. Along with this a person's nationality would also be handled differently today regardless of what type of format he would be broadcasting with. Unfortunately, during the times of Vere John and others the darker complexions of races, were called 'Colored' (regardless!), and it is because of the colored category being placed upon him during his days in radio, that historically, he is a part of the black history of radio.

Quickly, the role of the disc jockey developed into an evolution as another area began to develop it's own colorful station in the city of Baltimore, Maryland. Baltimore started it's own entourage of black broadcasters, and they also went a little further and added 'people of color' to all sectors of the industry. Local Headlines were highlighted with names that are still mentioned today, for example: the voice of the mighty Joe Bostic.

JOE BOSTIC
Example of Joe Bostic's Station's Listings

Entertainers Boosting Ranks of Baltimore's Own Radio Colony
March 26, 1932
All Departments of Radio Field Filled by Race
Announcer:
WCBM—Joseph Bostic
Singers:
WCBM-
James Coles, Clifton Spriggs, Reuben Parker, Rameses Rhythm Boys, and Bertha Idaho.
WCAO-
Chuck Richardson
WEBR
Buck Barnes, and Leon Nelson Pianists.
WEBB-

Rivers Chambers
Ambrose Smith
WCBM-
Tom Delaney, Jerome Washington, Godfrey Harris,
Orchestras: Ike Dixon, Percy Glascoe, Irvin Hughes, and Johnny Christian.

Looking at the roster, one can truly see that the city of Baltimore quickly, built up a radio colony of it's own. Increasing numbers of local entertainers began their broadcasting careers with gratifying results from their fans.

In order to understand the above roster and the pages to follow, one must be aware of the fact that artists didn't bring records to the station and leave them there for airplay, they sang their songs and played their music. Radio was LIVE and the excellence of it all was that it worked! No time for remixes, no time for edit, just pure listening pleasure with a touch of quality.

This quick influence of radio broadcasting and ownership was becoming popular within the community as they began to turn the dial from one station to another. The popular duo composed of Rivers Chambers and Ambrose Smith began to introduce their novelty form of entertainment billed as the Twenty Harlem Fingers, over WEBR on Saturday nights. Soon after others began to join the ranks such as the known groups such as Chuck Richardson, Jerome Washington, Godfrey Harris, the Rameses Rhythm Boys and others who were well known by local tuners in.

The many Baltimore piano pluggers were beginning to have the best performers appearing in their local stations. The hot performances of Harris and Washington were featured over the Keith station for more than two years. Both Chambers and Smith, the latest additions were well known in local and national movie circles.

Chambers was quickly recognized because of his families' involvement in music. Born as a prodigy of a father who was a musician in the Civil War and a mother, (Mrs. Alice Chambers) one of the pioneer music teachers of the city. In addition, his brother, Ulysses Chambers was former supervisor of music in the public schools and later concert organist at the Regal Theatre in Chicago.

On the other hand Smith had been an orchestra leader, arranger and composer. He had traveled extensively in America and abroad. He was at one time associated with Wil Marion Cook here and in Europe. The perfect synchronization obtained by these two artists in their broadcast was an excellent evidence of their mastery of their instruments. Together on Saturday nights they performed music by Leon Nelson, guitarist and tenor and Buck Barnes, popular baritone and musician.

GORDON HEATH

While Joe Bostic and his entourage were developing their own styles, the local newspaper headlines of "WMCA Scores Again" were being made as they highlighted a story telling everyone that a Negro was hired as a staff announcer. Known as Gordon Heath, this talented young actor, started work March 4th.

This was the third time in a week that WMCA earned the praise of liberals. The week before it had won two awards-one from the Schomburg Collection of the New York Public Library, the other from the National Conference of Christians and Jews-for its work in promoting interfaith and interracial harmony, particularly with its notable series New World A-Coming.

Heath, recognized the fact that he wasn't the first Negro to be a radio announcer-in fact, he himself reports he has been one on WLIB and WNYC as he says, "to the best of my knowledge, I hope I'm wrong, WMCA is the largest station yet to break radio's Jim Crow tradition."

(And of course, he was a wee bit wrong because as we have come to understand, Joe Bostic had done just as much or more in Maryland).

Heath, was a native New Yorker, born in 1918, and attended the Ethical Culture-Society School, the High School of Commerce and City College, and studied drama at the New Theater School. He announced, acted, wrote, and sang on the NYA (National Youth Administration) projects over WNYC; wrote scripts for and acted in the I'm Your Neighbor series on WNYC; narrated and acted on the Ave Maria Hour over WMCA; was narrator for the Pearl Primus show at the Belasco during their fall season; appeared in two recent television programs, on the CBS station WCBW (one program aired during the month of February he was heard as Abraham Lincoln, Lincoln's stepfather, and Walt Whitman). As a co-director of and actor in Owen Dodson's play, Garden of Time, he continued his career while rehearsing at the American Negro Theater, and in between times he was a scenario writer at the American Film Center.

With the various styles of Gordon Heath's broadcasting being recognized, this wasn't enough for him as he was also beginning to receive recognition for his acting abilities. With the control of the neighborhood beginning to develop the Disc Jockey was beginning to permeate the communities which they served.

ROY WOOD

Roy Wood's broadcasting career began in 1948 as a radio announcer. Almost, immediately, he had taken the community ears away from other stations as he informed them of the daily news the way he knew they wanted to hear it. Informatively, he delivered the community news the way he knew they would appreciate it.

Roy Wood developed his image and career path with his take charge type of personality. During each career move made by Roy Wood he was able to impress the management of WVON, WHFC, WENN, and WGAT which allowed him the opportunity to tell the stories to the community the way they wanted them to be told.

Roy's career advanced, his style of broadcasting news and commentaries began to reach others across the airwaves as he moved swiftly with the times. However, the highlight of his career can be remembered by many when he visited for one full month in Southeast Asia and took the opportunity to interview many of the Chicagoans that were fighting in Vietnam. (Which also has been reported that as many as 10,000 phone calls from servicemen relatives were made to WVON requesting that he talked to their loved ones). Roy Woods= sports/talk show was also one remembered by Chicagoans as they listened eagerly to hear his voice and the voice of his peer Ernie Banks during their 1958 broadcasting relationship.

Roy Wood's career is also synonimous to that of Don Cornelius of 'Soul Train', as he mentored Don Cornelius and helped him to get 'Soul Train' started at WCIU TV before the dance program became a national one.

In the 1960's he developed a news show entitled, "A Black's View of the News", at WCIU. And later helped to establish and to become a top executive at National Black Network News in New York City. After spending several years in this position, he later moved to Birmingham, Alabama and held the position of news director at WENN-WAGT radio until his death.

Roy Wood, a legend in his own right passed away in October of

1995; but, his gracious way of telling the news from a perspective inwhich the everyday person would understand it, can long be remembered by many as the voice that would always end his broadcast with . . . "Now . . . Run and Tell That!" As he opened doors for others to get the opportunity to come through.

RICHARD STAMZ

How does one become the "Crown Prince of Disc Jockeys?" Richards Stamz tells you how and why as he reflects back on DeKalb, Illinois. It was there where the city held a disc jockey popularity contest where a collection barrel for the then current 35,000 population and recognized him as number one disc jockey as they crowned him in front of the people that voted for him. Richard still awes at the very thought that he didn't think the people would recognize him as the normal opinion had usually been geared to the white disc jockey during this particular time.

Working under the auspices of Jack Cooper, Richard Stamz at the youthful age of 94 has wonderful memories to recall of his days in broadcasting. Richard Stamz, known as *(OPEN THE DOOR RICHARD)* is in high demand for speaking engagements at institutions such as Michigan State, and today makes his goal to let the young people know what his career has been all about. Richard still enjoys reviewing the historical files he posesses of disc jockeys, such as, Ken Knight when he started the first black gospel broadcast on CBS which broadcasted to more than one city in the southern region. Many of his other files from the Chicago station where he worked with Jack Cooper as they worked with the black baseball teams broadcasting their games play by play are still bringing tears of joy to his remembrance of a time of hard work and dedication.

Before arriving to Chicago Richard remained true to his home town of Memphis, Tennessee and after graduation from high school, he attended Lemonyes Owens College for three years and Columbia School of Broadcasting.

Richard's career has been an enormous one. Richard's start in broadcasting began before he signed on with WGES, as he was apprenticed by the well known Jack Cooper. After his training he began his professional broadcasting career at WGES 1390 AM in Chicago, Illinois from the year of 1949, and remained there until 1965. Richard'show was entitled, "Open The Door Richard". As his career grew he achieved the position of an On Air Performer and Salesman at WVON 1390 AM, in Chicago from the time period of 1965-1967. Later, he became National Sales and Artist Performer for Columbia Recording company from the period of 1960 to 1971.

But Richard Stamz can also be remembered from his hard working days of Directing and Producing, as he took the title as Host of WBKB TV-7 on January 21, 1956 from 11:30 to midnight and began hosting the "Richard's Open Door" television program. And has been noted as the first African American to have ever host a television program on a major network.

Later Richard Stamz became an organizer and board member of (NATRA) the National Association of Television and Radio Announcers from the years of 1952-1961.

As a result of his hard work throughout the years he has been named "A Legend" by the R&B Record Division of America because of his ability to break the hit, entitled, "The Green Door, What's Behind The Green Door" by Jimmie Reed. As Richard admits to naming his once owned tavern "The Green Door".

But this story could not be closed without Richard Stamz telling of some of his stories and triumphs.

During the late 1930's and early 1940's he and Jack Cooper drove through the black neighborhoods promoting the old Black basketball team, "The Chicago Giants". As they promoted them they used a truck with a speaker to announce the games to the neighborhood. As a result people would prepare to come to the games and those in the community lived by this method of promotion which was also fun for the disc jockeys to do during the early days of baseball and promotions. Promotions for advertisers could be done for all of the station's sponsors and in later years they would hook into the radio stations for remote broadcasting. Richard explains that this is actually how he was able to get in the radio business. And admits that this was fun and it also allowed him a lot of room for creativity.

Although, Jack Cooper has been the one credited to his being on radio in early years, Richard admits that there was also another person who helped him attain his career goals, and that was Ann Southern, (the then famous white actress), who he traveled with for three years and did some movie parts.

As time went on Richard's reputation was building more and more and finally he was hired at WGES 1390 AM. The executives at WGES realized the potential of the Black consumer market and that a "Negro oriented format" would be an appropriate medium. And as Richard admits, "was it ever, profitable and successful!"

Judge Edith Sampson the 1st black female Judge in Chicago was on his first ever broadcasted television show.

On WGES Richard Stamz title of "Open The Door Richard" was the name used because he moved fast and so did his words. There never seemed to be enough for him to do. He did his daily radio stint, and sold air time. He went to great lengths to find "specials" and unreleased recordings of great musical stars. At this time he was also operating his sound truck business, because, he was also in high demand for emceeing talent shows.

As he developed his career more and more he was able to make WGES radio station one of the most popular Black oriented radio stations in America during the 1950's and 1960's. His ability to attract advertisers and promote their products was widely known. Richard admits that this wasn't all done alone, as he names other greats who worked with him during his career. Radio greats such as Al Benson, Sid McCoy, and other WGES greats helped shape and transform Black radio into an important medium to reach Black consumers and set the music pattern for America's Black population.

One story which Richard tells of his career whole heartedly is the time he had to use the power of the airwaves to get racist police off of his back.

As Richard tells it . . .

"Once while working in radio I was coming home from work in my specially built 8 centimeters and 6 foot forward beautiful dodge convertible and the police was following me with their spot lights blinding my vision. I then pulled over and told them to get that light out of my face. They then proceeded to ask me for my license. I refused and was told as a result that I would have to give them a cash bond at the police station if I refuse to pay them cash in their car. They said, well, Richard we'll have to take you down to the station. When we got to the station I gave them a $1,000 bill and they told me that they couldn't change a $1,000 bill for me. I told them that this wasn't my responsibility and that they had gotten American Currency for a cash bond. And then they told me that they would have to go to central headquarters and see if they could change it. I told them they would have to give me a receipt for a $1,000 bond, and about two hours later they called me back and told me that they had my change. While at the station, the police captain looked at the officers and said, "what are you supposed to do?" they proceeded to do a little song and dance and recited, "go out and search crime". He said to them, ok go on and get out of here. As they left the white police captain told me that I was guilty and there was nothing he could do about it. I said ok and left, even though I still felt that I hadn't done anything wrong. The next morning I got on the radio and told others in the community so they would know that they had a cash register in the back seat of their police car. The following morning the captain came and wanted to come into the studio. I told him that this was under the auspices of the FCC and that he would have to call the FCC. Coincidentally, the owner of the station was there and he called the FBI, and the FBI called the police captain upstairs and told him that he couldn't come into the radio station without consulting the FCC if he wanted to demand anything. He left very mad ad the next day the Chicago Police Department Internal Division came by and asked me, "Mr. Stamz, what is this all about, we're here about a cash register in the back of the squad car?". He said, "Is that real?" I said, well these guys took a few dollars from me and they rang up a few dollars from me in the back of the squad car. They proceeded to ask me if I wanted to make a statement, I told them oh! no! I just want to tell the truth about this. So they left. And for four straight mornings I got on the air and made the statement that this squad car had a cash register in the back seat of it. Once they realized the power of what I could do with my position at the station, they decided to take another route. On the fifth day I got up to go to work and there was an envelope under the door with five hundred dollars in it. They were in one hundred dollar bill denominations. I knew what it was, it came from the police department. The state did everything to make me stop, and they couldn't make me stop. I took that same five hundred dollars and joined the NAACP for life. That's my story", laughs Richard, as he smiles to reflect back on such a historical time in his life.

These were great years for Richard Stamz and his other peers in the industry. They got the opportunity to use their influence on the artist's career and upon the listeners and admits it still all seems as something unbelievable. Everybody tuned in to hear their broadcasts because they were in touch with the celebrities and the community.

And as time would tell Richard Stamz's great success paid off with his "Richard's Open Door" television program which featured music, celebrities, talk and entertainers. As a result of his popularity as a radio and television personality, his name was widely heard throughout the country from those in the industry, as he earned the name "Crown Prince of Disc Jockeys". He is recognized as a pioneer and an innovator. Some thought he was ahead of his time because of limited opportunities for Blacks in the media; but, his colorful personality attracted many national and local sponsors for his radio and television programs.

Richard Stamz produced blues and gospel artists, produced radio jingles, and worked for Columbia Records, but, he admits, that he couldn't do all of this and not think about others. In 1952, he and another radio personality "Jack The Rapper", along with several others held a meeting to organize Black radio and television stars. Stamz served as chairman of the Executive Board until 1961.

Richard also made sure that he worked with the community as he gave a great deal of his time and resources to the Boy's Club and the YMCA. As a community and political activist he worked to help improve the quality of life for Chicago's Black community. As his numerous professional, civic, community awards and honors reflect this commitment.

And today, Richard Stamz at the young age of ninety four is still active as ever as he lectures at colleges and for small groups.

CHAPTER THREE

WILLIAM BLEVINE

The South's not far behind . . .

As we move along our journey we can see that this was starting to become a very popular career for our race to become a part of. Not only were the northern cities acknowledging the ability of our people to become announcers, but, it was also beginning to ring true in the south.

Birmingham, Alabama began it's acknowledgement by hiring and promoting the abilities of William Blevine. Blevine's broadcast was known as the 'All-Negro Radio Hour' and was recognized as such on March 3, 1936 in the Washington Tribune.

As the "All-Negro Radio Hour" was beginning there was becoming an increasingly cemented relationship between the colored and whites of the community. This was all attributed to the first Negro to be featured over WSGN, William Blevine.

"Our Gospel Singer," program with William Blevine, was broadcasted each Tuesday and Thursday night and on Sunday at 1 p.m. and was publicized as a goodwill gesture by the Birmingham News and Age Herald.

In addition to these weekly features, Blevine made it possible for many Negro leaders by sponsoring civic projects to speak to thousands throughout the southeast over the radio. He also bought prominent Negro choirs from all over the district and state.

Just as some of the northern cities assisted their communities by involving participation of their communities, William Blevine was doing it for the south.

Shortly after Blevine's rise in radio began in the south, the city of Cleveland was beginning to make it's marking with Rev. Glenn T. Settle in 1938. As pastor of the Gethsemane Baptist Church of Cleveland he met with program director Worth Kramer of Columbia Broadcasting System and requested that his church had radio time for his choir to perform. As the program was initiated and developed it later received national attention and on Jan. 9, 1938, the "Wings Over Jordan" broadcast became a part of the Columbia network which could be heard locally over radio station WGAR. Rev. Settle and his choir began to tour locally and eventually nationally as they attracted as many as 15,000 people at one individual concert. And those who never got a chance to see his choir perform in concert were able to hear the broadcast as it aired over fifty CBS stations on Sunday mornings at 10:30am.

It is because of the Jack Cooper's, William Blevine's, Richard Stamz's, and ham operators of this time period that another Disc Jockey arose in 1939, who's still on the radio today. This Disc Jockey is the well know Hal Jackson out of New York City. In a recent newspaper interview, Hal Jackson spoke about working in the earlier days of broadcasting, as he told it to one USA Today reporter.

HAL JACKSON

Hal Jackson's fifty years of broadcasting has pronounced him to his rightful status when he was honored at the New York's Apollo Theater by stars including the Temptations and Bobby Brown, plus video tributes from Michael Jackson and Stevie Wonder.

And with all of the knowledge of his knowledge of black radio's past he explained when asked, What's ahead for black radio? Honestly, Hal Jackson acknowledges "its past".

Of course, there are many people who wouldn't understand his statement, and would probably think that he's trying not to give up what he's had a lifetime of. Unfortunately, this is far from true, Black Radio's past is what held us together as a community. Black Radio kept us informed of what we needed to know about and true Black Radio cared about the people it served. Not the record companies, not the promoters; but, the people. The type of music, the local news from the black community that the people wanted to hear is what should be ahead for Black Radio. Instead, technology, and greed are replacing this and Black Radio isn't Black Radio when you do this. Black Radio is for the people and people deserve something good to listen to is what can be found in Hal Jackson's answer.

Hal Jackson explained during his interview, "You listen to a lot of music of the '60's, it all had such a good base." He also stated, "That's why everybody's going back to grab these songs; the songwriting is better." And Jackson doesn't speak of Black Radio because he sees it as a dying dinosaur; but, because of his experience of spinning records since 1939 when he started his career at WINX in Washington, DC.

Today, Hal Jackson is producer for TV's Showtime at the Apollo and hosts Sunday Morning Classics on New York's WBLS FM, attracting 100,000 listeners every hour. He has been and can still be heard from the hours of (8a.m.-2p.m.) as he has for the past three years.

But, Hal Jackson's career doesn't stop here as he's group chairman of Inner City Broadcasting, which owns five radio stations and the Apollo Theatre.

Noted as a prominent figure in black music, he's given several acts their first breaks, including the Jackson 5 in 1968. As he recalls, "I remember when (Motown founder) Berry Gordy called and told me that he just signed this group. . . . I was doing a Miss Black America pageant at Madison Square Garden, and he asked me if I could put them on the show. I worked them in with Stevie Wonder and Curtis Mayfield."

And admits that their quality and style of music makes him still apt to play a Jackson 5 oldie as he leads the retro-movement, which he says will grow among black stations. "When I mix hits with older things, the listeners love it. Especially young people."

"I may use Anita Baker in a segment, followed by Count Basie, then come back with a group thing like New Edition. It's the flow of the music that holds the audience."

"We're going to hear more and more of the '60's music being re-recorded," he adds, citing Sybil's remake of Dionne Warwick's '63 hit Don't Make Me Over, No. 3 now on Billboard's dance chart and No. 16 on the black singles chart.

Hal Jackson strongly feels the public is growing weary of new jack swing (a mix of rap beats and R&B singing) dominating black radio. In its place, he says, the blues is gaining new popularity.

"I never do a Sunday segment without the blues. . . . I'm seeing young people checking into artists like Bobby Blue Bland. I go back and bring up some Etta James. There's people like Big Joe Turner, who I will never let people forget," admits Hal Jackson as he tells some of the reasons why he keeps the past in his present day format.

Hal Jackson still shares his memories with us and brings back some of the oldies but goodies to share with those of us who can still appreciate it. There's more to tell of Hal Jackson as others tell the rest of the story when we read about the 'Bad Boys'. But in order to understand what he and others had to operate with in the earlier days of broadcasting can be defined in the following examples of their styles of personality broadcasting. Imagine hearing the strong personality voice of Hal Jackson telling this type of story to those in his neighboring community.

An Example of Radio Scripts Disc Jockeys Used Written in the 40's
FEDERAL SECURITY AGENCY
U.S. Office of Education
Washington, D.C.   Release AM Nov. 17

Joe Louis on "Freedom's People" Nov. 23

The first name among Negroes in sports—heavyweight boxing champion Joe Louis—will be featured in the third broadcast of the "Freedom's People" series over the National Broadcasting Company's Red network from 12:30 to 1 p.m. EST November 23.

"Freedom's People," a once-a-month series sponsored by a national advisory committee with which the U.S. Office of Education cooperates, dramatizes Negro contributions to American life. The November program features Negroes in sports.

Jesse Owens, who showed body-worshipping Nazis something in the way of physical ability when he captured three first places in the 1938 Olympics at Berlin, will join Louis on the broadcast. Bill Stern, race sports commentator will interview Owens; Ken Carpenter will question the taciturn Louis.

Owens holds world records in the 100-yard and 220-yard dashes and the running broad jump. He won the 100 and 200 meter and broad jump events at Berlin and also ran on the American team which won the 400-meter relay. After a professional tour and various other jobs, Owens has returned to Ohio State University to complete his studies.

Louis, undefeated in the ring since he won the heavyweight title in 1937, will probably speak from California where he is residing while awaiting a call from his draft board.

Cab Calloway, whose "hi-de-ho" has been a jazz trademark for years, will lead his orchestra in one of the musical numbers. On the more classical side the broadcast will present the Golden Gate Quartet, well-known to radio and concert stage audiences.

The "Freedom's People" series is made possible by grants from the Rosenwald Fund and the Southern Education Foundation, together with broadcasting facilities offered by NBC and network stations.

Acting for the U.S. Office of Education in cooperation with the special committee are Dr. Ambrose Caliver, Senior Specialist in Negro Education, and William D. Boutwell, Chief of the Radio Service. Irve Tunick, veteran radio author, writes the scripts. Original music & arrangements are prepared by Dr. Charles Cooke.

CLIFFORD BURDETTE

Another rising star in radio who's popularity may have been quieted until now is that of Clifford Burdette. Clifford Burdette wanted to work in the field of broadcasting so bad at the age of 27, and he finally accomplished it by working in a silk store eight hours a day as a salesman to support himself, and his broadcasting career. Clifford Burdette managed to pay for his career development out of his pocket after funds ran out from other sources. As he broadcasted his program, "Those Who Have Made Good", he kept in mind the promises he made back home in Atlanta, Georgia to his family and friends. He developed a style of programming which recognized those of his race who had or were beginning to accomplish their goals as he highlighted them and their careers on his broadcast at radio station WNYC. Along with this defeat he managed to receive full support from the NAACP as they paid for the first 10 weeks of expenses for his program in order to keep his broadcast alive and well.

Clifford Burnette regards his management of the airwaves as a result of his roots, stating, "My parents were poor, but, they were determined that I should have an advanced education, so that I could help others of my race today and in future generations. They implanted that idea in my mind, and I have never forgotten their ambitions for me".

As he started out as any other teen in the south he decided to do a little more in the area of the arts by performing with the school glee club, and traveled over the state reciting poems based on the Negroes of the South. At eight years old he bought his own radio set by selling newspapers. Determination was instilled in Clifford at an early age and as a result WNYC gained recognition for his talent.

An example of the power of his broadcast can be found in one story written about him in 1941 which highlighted his "Those Who Have Made Good" broadcast which gave the following account using 30 minutes of air time to highlight notable Negro gains of the day.

The Negro In 1941
WNYC's Those Who Have Made Good

1. A Negro Army Air Squadron was formed, and volunteers called for.
2. The Supreme Court ordered equal facilities for whites and blacks in railway trains.
3. Samuel J. Battle, NYC's first Negro police official became the city's Parole Commissioner.
4. The U.S. Naval Academy shelved its Jim Crow policy in athletics, playing against a Cornell football team which had Negro Sam Pierce in the backfield.
5. The Rev. Adam Clayton Powell, Jr., won Harlem's main seat in the N.Y. City Council.
6. Joe Louis kayoed four challenges for his heavyweight crown, and now is prepared to face all corners as a Class 1-A selectee for the U.S. armed forces.

"And we can also say Clifford Burdette Made It Good!"

CHAPTER FOUR

NAT WILLIAMS

Problems! Problems! Problems!
WDIA Saved from Bankruptcy

Because of the astonishment of the new sounds and broadcast given by some of the previously mentioned disc jockeys, their existence was also a life saver for radio station WDIA which was going into bankruptcy. Through the visualization of Nat Williams WDIA changed it's programming format and saved the station at the same time.

The headline was given as follows:

Blues, Bebop saved WDIA from bankruptcy.
Negro voices influence sales.

A Memphis school teacher by the name of Nat Williams proved to be a god send to Bert Ferguson and John Pepper, white owners of WDIA. They were on the verge of bankruptcy. It was hard for them to out do the competition of other local networks.

The jovial Williams solved their problems when he convinced them that they should switch from hill-billy music and beam to the neglected Negro market. As Nat Williams displayed many facts and figures showing that Negroes purchased at least 40 percent of all merchandise sold in Memphis, they decided that this is the avenue they needed to take in order to survive in the current market.

When they put "Nat Dee" on the air as an "early bird" disc jockey, (which stood for morning hours); radio advertisers paid attention as he did his early sunrise show `Tan Town Coffee' and his evening program called `Tan Town Jamboree'. Before long their red ink changed to black and they added other Negro features.

Immediately, Memphis 205,000 Negroes began to listen, as their mail indicated that programs were heard regularly in Ark., Miss., and Ala. A special Hooper survey was taken and showed that WDIA had 69.7 per cent of the day time audience, white and colored.

WDIA's programming offered such interesting shows as Willa Monroe's hints to housewives; William's daily shot and his Sunday forums. A disc jockey who will be later discussed, Gatemouth Moore had his soul-inspiring religious hour and bopster Hot Rod Hulbert hosted the 'Sweet Talkin Time' program for teenage listeners. Soon after, LeMoyne College was waiting for it's application approval from the FCC for a station; and if this were allowed Memphis will be in for a dose of cultural shows by its students. The white announcers always referred to the Negro women as "Miss and Mrs." And it was revealed that a Negro's rendition of a commercial had a greater appeal to Negro buyers of his sponsor's products.

The widespread popularity of the station was growing more and more as the owners put more stock into the station and into the community. They finally received more control of the airwaves and later became a twenty four hour radio station and thanked the community by supporting it strongly.

They began to put even more emphasis on the community as they recognized those who worked long hours in the cotton fields. The very thought of bringing broadcasting to the then known negro community elated everyone. One noted recognition was that of 71 year old George Carter who was finally able to hear what he had only previously been able to listen to in his home, and now at work he was able to listen with the radio which he won from the station.

MAURICE HULBERT
Hot Rod

Acclaimed in the National Hall of Fame. On July 30, 1926 Hot Rod was born. Hot Rod studied tap dancing and became the leader of his own band. Leaving Arkansas to Memphis, Tenn. he began to work at WDIA. White Dee Jay Buddy Dean compliments Hot Rod energy and talent and recommended him for Baltimore's popular station. When the station saw that they had a large population of black listeners they were not appealing to and they decided to add a black dee jay to their payroll. Hot Rod left the all black station WDIA to come over to WITH. Hot Rod's words were awesome to listen to and has been recognized by Hall of Fame Baseball broadcaster Chuck Thompson as someone who was fun to listen to, and very bright. Chuck admired Hot Rod when he used terms such as a "Dilly-A-Blah", which meant you get on the phone and you blab a little bit. In the mid 50's he said that 1/3 music, 1/3 commercials and 1/3 me is what you need to make a broadcast work. Which became a very successful formula.

Hot Rod had an opening broadcast of his ride into space on his fabled rocket ship. Other dee jays such as Jack Gayle enjoyed listening to Hot Rod when he was on the air or emceeing. Once, Hot Rod was at a show where the Orioles were performing and Hot Rod would perform with his rocket ship show, the other dee jays would just stand in awe admits Jack. Jack would do his 'hound dog show' and would come in to admire Hot Rod as he performed his show which came on before his.

In 1954 the advertisers such as Gunther Beer took out a 2 week sponsorship on Hot Rods show. Because of the way he excite the audience with his Dilly-A-Blah and Voo-sah coined phrases.

Because of the payola scandals of the early 60's Hot Rod's career took a change because a tip sheet type of newsletter he had which was funded by record companies in the advertisement of the newsletters. Hot Rod had to leave WITH and go to Philadelphia to station WHAT.

By 1961, Hot Rod was heard on WHAT in Philadelphia, WWRL in New York and WWIN in Baltimore. In 1963, Hot Rod negotiated with Jake Emery and came back to Baltimore's WITH. At this time WITH was facing a new type of music called soul music. WITH was sold when this new change in music came about and the new ownership released Buddy Deane in September of 1964 of his duties and changes were taking place all over the country.

WWIN's Personalities such as Long Lean Larry Dean, Tall Fat Daddy Johnson, Kelson Chop Chop Fisher hit the airwaves and it became a natural fit for Hot Rod. Sam Beasley, had a teenage show and recalls how Hot Rod was such an entrepreneur. Sam remembers the record hops, boat rides, and things of this nature which he observed Hot Rod doing. Sam calls Hot Rod a master of the trade.

Known football player, Lennie Moore who later became a disc

jockey recalls how Hot Rod was recognized as a great personality on the air, but, he was even more awesome when he emceed programs. Lennie looked at him as an overall entertainer.

Hot Rod began staging shows in 1954. Once he arrived to a show while riding on a donkey. This drove the crowd wild. Another time, Hot Rod came to a show in a helicopter in his space suit with models dressed in space suits and drew a strong crowd as a result of this.

Mary Claiborne, was public affairs director at WWIN, and recalls how Hot Rod helped her when she wanted to interview a well know female vocalist for the station who had decided that she would only be interviewed by television personnel, because they had cameras everywhere. After interviewing with the local television stations this star said that she sorry that she wasn't going to talk to the radio PR people. Mary informed of how upset she was and told Hot Rod that the station played her records, and she came to town and she would not talk to the people on radio. Hot Rod called her after he found out what hotel she was staying in and he called and told her that when she wanted her record played to take it TV, if you want radio to help you, then you give us an interview. Instantly, this star got in the car when the chauffeur arrived to her hotel to pick her up and she came to the station and gave a very long interview.

During the time period of Muhammad Ali's fighting the issue of not wanting to go into the military, Hot Rod, told him this phrase, "If I tell you a mosquito can pull a plow, don't ask me how, hitch him up, and ride him to the moon".

Joe Parker recalls how Hot Rod gave a boost to his program at the local cocktail lounge. At the civic center when stars such as Moms Mabley, the Temptations, were there Hot Rod was there with them. Many times he would go on the bus and train tours with them as he would emcee their shows on the road while taking some of the Maryland talent with him.

Al Jefferson who was once program director at WINN recalls how much he admired Hot Rod as a legend; because of his out of the world personality type.

Goo-ga-mooga was a phrase he coined, meaning awesome. Even the Temptations used this phrase in their song, "Ball of Confusion". By the 1980's Hot Rod married Brenda Hulbert, and became manager of station WBGR and WEBB. He and his staff made these stations two of the most successful gospel and rhythm and blues stations in the country.

In 1993, he retired at the age of 77. Soon, afterwards, he became ill and began to carry a note pad with him after losing his voice to illness.

Buddy Dean said Hot Rod called him when he had booked a show, and needed some money to pay some people off for the show. Buddy recalls that Christmas, Hot Rod mailed every penny back to him.

In the early 1990's in Cleveland, Ohio Hot Rod was voted into the Rock 'n' Roll Hall of Fame.

Disc Jockey, Randi Dennis of WERQ, explains that when he came to Baltimore in the early 70's, he didn't know anyone there, and tells how Hot Rod took him in and set him straight. Randi explains how Hot Rod became a father figure and a friend to him.

It was said that Hot Rod was in the fast lane of radio and that no one could keep up with him. Sir Johnny O, knew Hot Rod when he was 16 years old and recalls how Hot Rod told him that he would let him be his co-pilot, and felt that this was quite a thrill.

Hot Rod's 80th birthday was celebrated on July 30, 1996, many friends and family, and previous associates came to honor his many years of broadcasting. Dec. 24, 1996, Hot Rod passed away a smile was in his heart. The Hot Rod story 1230 am WITH, Baltimore.

PAUL E.X. BROWN

There's one Disc Jockey who's quite familiar with stations giving away raffle tickets and finally letting the black race get into the industry and that's the powerful Disc Jockey named, Paul E.X. Brown. After growing up in Mississippi and graduating from High Sumner High School in St. Louis in 1929, Paul E.X. Brown decided to further his education by attending the University of Minnesota, Columbia College and Moody Bible Institute. Paul E.X. Brown still recalls that if it hadn't been for the earlier years of the Depression Era, that his career in broadcasting may not have taken off as fast as it did.

Initially, Paul E.X. Brown worked as an editor of a 16 page weekly called the Chicago World, which, included 15 minute shows that were sponsored on WJOV. As he recalls hearing the broadcast of Jack Cooper in Chicago and recognizes the power of information that was being given to the community as a result of Jack Cooper's efforts. A little later he recalls that he had no idea that he would be getting a job at the station when he was approached by Eddie Ernester. As Paul E.X. Brown was asked to do a fifteen minute show, he didn't know exactly what he was going to do, or say, it was from the only knowledge that he had, 'writing scripts' that he used as his working tool to make it through his first broadcast at the station. As he tells the story, he tells about the fact that they wore white shirts a lot to work during this time and he put on his white shirt and tie and went in to do his first broadcast at WJOV. After the show his shirt was so wet from perspiration, he still laughs today, saying, he could have wrung it out with all of the water in it. Later, the next day, he was informed that management wanted to see him, and he felt as if he may have done something wrong with his broadcast. (Remember, this was a time when the FCC was steadily watching stations for breaking rules, and the military was making sure that they didn't broadcast any codes to the enemy), but, instead he was offered employment, as the manager informed him that he listened to his broadcast while playing on the golf course, and was impressed with what he had heard. Paul E.X. Brown was excited over getting this position and began to do the daily record show. This became a record high for the Lake County area of Illinois, because, it helped a community that was living through a depression.

During this time of depression the Defense Program of World War II made the station send in a list of all the songs which were going to be played as they were given to the record librarian to record, and once the list was made, the Disc Jockeys at the station had to play from the list, without changing any one song. Paul E.X. Brown recalls how the effect of what they broadcasted had a lot to do with industries such as the railroad transportation department, and factory makers such as those at steel mills which also had to be protected. Paul turned in his program to the librarian entitled, 'Headlines In Review'. The tapes that were made at the station were recorded inside and out and the engineer's responsibility was to format them for air play, and as said before, the material could not be changed once submitted.

After getting some years of experience, Paul E.X. Brown decided to move to the growing city of Atlanta, Georgia in January of 1947. And later became Announcer and next Program Director at WERD from 1950-1955. It was during this time that John Wesley Dobbs, the grandfather of Maynard Jackson and others such as the brother of Zilla Mays were trying to do something for the kids in the community. Later, they began a program at Magnolia on Sunset for the teens and to their surprise this successful event also bought in whites to the black community. Paul remembers that he too, wanted to be a part of this growing area and began to write to several stations for employment. Before all of this had occurred, Paul E.X. Brown recalls that he wrote to all the stations in the area, which there were only five, and was told by the only one response he received back—"No Way!" Later in 1948 he was able to work at the station WEAS in Decatur, (now known as WGUN), and soon met with John Hardin who had been following his activities and as a result received air time by the sponsorship of Mr. Hardin. Soon after he was able to become employed by WEDR in Birmingham, Alabama from 1949-1950 and greatly recalls working with the unforgettable Shelly Stewart.

Once, Paul E.X. Brown had the honor of broadcasting a game played by the Atlanta Black Crackers baseball team as they competed against the Brooklyn Dodgers, and he laughs as he recalls that each time he called the team 'The Crackers', instead of the Atlanta Black Crackers he would get hit over the head by the station manager. It was because of the appeal of the broadcast and play by play that WEAS was then allowed to broadcast the game all summer.

Soon after, there began discussion about station ownership and a wealthy white man from Chattanooga, Tennessee came to see the black wealth in Atlanta he had heard so much about. He began to pull together seven members of the community to meet with to discuss the black ownership idea, and meetings were held with J.B. Blayton, John Wesley Dobbs, A.D. Walden, Yates, Ed Medlin, and two other gentlemen and as they met at Trust Company and tried to get everything together, the word was out in the industry that a small country station was beginning to have financial problems. This station contacted J.B. Blayton and a deal was cut for them to buy the station. But as a technicality history would tell you that it was sold on the front steps of the courthouse after bankruptcy, because during this time, whites wanted to protect their image. After the contract was finalized, J.B. Blayton had to assure the owners that he would not move into the same building because the building management didn't have separate restrooms and they wouldn't be able to accommodate the negro if they worked in the building.

As the station moved, opened and developed, Paul E.X. Brown wanted it also to be recognized that there was one white Disc Jockey that remained with the station and always throughout his career was dedicated to the black listeners and never did he let white influences affect his decision of working at an all black station, and that Disc Jockey is Bob Brisandine. Paul calls Bob, 'One of the Boys' and remembers his "Bouncing With Bobby" broadcast. Paul also recalls another white disc jockey, whose family thought that it wasn't right for him to get his pay from a negro, and as a result, out of the two whites at the station one stayed and one left the historical station of WERD.

As Paul E.X. Brown left WEAS and began working at WERD, he remembers asking Ken Knight to let him freelance his "Hi Neighbor" show, and was given the opportunity to do this on WERD and at night on WATL.

In 1953 after Ken Knight left the position of Program Director, Paul E.X. Brown became the Program Director and remained in the position until 1955, when he left to work at the Pittsburgh Courier, in their Atlanta office. In 1959, he went over to radio station WAOK and worked with Public Affairs until 1962, when he left radio and began employment at the Atlanta Coca-Cola Bottling Company as a Sales Manager until his retirement in 1975.

CHAPTER FIVE

CASANOVA JONES

The best way to introduce Ken Knight is to have one of the industry professionals who had the experience to work with him; i.e. Casanova E. Jones, Jr. . Casanova Jones started in radio in Springfield, Massachusetts. As he explains:

"It all started in more a sabbatical type of entrance into the industry".

While recalling the fact that he was asked to represent a young lady who wanted to appear at a local talent show, he was more than anxious to help when he found out the winner would get a chance to appear at the Apollo Show in New York. Quickly, he took a chance of becoming her manager so she could get the opportunity to perform on stage. Casanova Jones took time away from his busy career to assist her in filing the proper papers to become her guardian. (During this time if a person was to become a manager which was a current law of the state. They would have to file to become a legal guardian). Under these conditions he managed her for a year. Later, Casanova enrolled back into school again and completed his education in radio in 1968. As he completed his career he went to his mother's hometown of Daytona Beach, Florida and heard that there was an opening at one of the stations there. This particular station had decided to change it's format and was a disappointment for Casanova as he moved back to his hometown of Jacksonville, Florida. When he arrived back home, his sister and aunt referred him to talk to Ken Knight who had the local television show. Casanova followed their advice and for at least seven months Casanova Jones found himself hounding and hounding "Mr. Knight", (as he calls him), and remembers the feeling he had deep down inside that he would now get the chance to become the hottest soul disc jockey to hit Jacksonville, Florida by finally receiving a position to work under the direction of Ken Knight. Casanova's new beginning wasn't as big of a challenge for him as it could have been; had he not gone to the station on his own time to learn the new techniques with the equipment. But, he felt that Ken Knight had something else in mind for him, as he recalls one particular night one of the disc jockeys wasn't able to make it in and he was informed that he would be doing the gospel show. At first Casanova thought about it with some anticipation. Recalling the whole ordeal, he remembers ho he thought that he could change the gospel trait of the area by using the same style and attitude as a soul jock when he did the gospel show. Quickly, he found out that he couldn't do this without receiving a lot of heat from the city of Jacksonville, who eventually accepted him. Listeners slowly began to accept him as he broadcasted from radio stations WERD and WCHY on Beaver Street behind the adult education school.

Soon after, he received a spot with Ken Knights' television show as a talent coordinator. While he worked busily at Ken Knights' office he remembers how he would set the appointments, time their acts and set them up for the show.

But, changes were ahead for Casanova Jones as he trained under the auspices of Adrian Kenneth Knight. A new radio station was being bought by Ben Tucker, and Billy Taylor along with a third partner in Savannah, Georgia. They had asked Scotty Andrews to work there, and he stayed there for two weeks only to go back to Atlanta. Cepio Collins, a well known friend of Ken Knight called him to recommend a replacement. The recommendation went to Casanova Jones. On the following Monday after finding out about the job he was offered the position at WSOK where he worked for three years. Later in 1976 he started working for WEAS-FM and has been there every since.

Casanova Jones admits that his lifetime dedication has been to those who have helped him to maintain his position. He also admits that the community has been the strong force which has been there for him throughout the years. And the community, recognized Casanova's efforts in radio as they honored him on Oct. 2, 1994, with an award from a local civic group for his dedication to the community. Casanova's beliefs of dedication and the community started in Jacksonville, Florida and has reached across state lines to give the community what they need to get from the radio when they turn it on to hear his voice.

"Radio has been a happy time for me", as he recalls the time when the station had a competition for the listening audience to pick the disc jockey who could be Santa Clause to the children. The mail came in groups but there was one influential letter from one little girl who wrote an inspirational letter to have him as her Santa. Early that Christmas morning he went to her home and gave her the Christmas she and her family would never forget. Another example could be found in the way he supported the community for Thanksgiving, when he picked out families that he thought needed support for

Thanksgiving and would knock on their door and give them donations for Thanksgiving dinner without them ever knowing of his coming to bring them a little joy into their lives and to enjoy the American Thanksgiving as others across the country would.

Once, Casanova was walking down the street and met W.W. Law who was one of the strong leaders in the city and was told that no one in the town knew who he was. When approached with this Casanova asked him to give him six weeks and they would. As a result, he came up with the Beautify Your Neighborhood competition which became a big community event for the residents of Savannah. Beautify Your Neighborhood also received support from the city with donations of trash bags and other necessities. The community worked hard to beautify their area. Casanova would then thank the community by giving them a disco that night and recalls the enjoyment of doing such a great community thing.

Casanova's moments in radio doesn't stop here as he proudly speaks of the chance he got to interview Chappy James, Billy Daniels and one of the most memorable ones is that of Alex Haley, who had just completed his book, "Roots". Haley's broadcast turned out to be a six hour radio broadcast, and now Casanova wishes that he had taped such a memorable event.

Closing out Casanova's career he admits that his career has been found to be quite enjoyable and admits that the experience to have worked with Ken Knight, and getting the chance to see him as the regular husband image at home also gave him the insight of the person who could leave business at the office. Another acknowledgement is to the great pioneer, Johnny Shaw—"the Devil's son in law!"

Casanova closes out his story for now; but, he can still be heard on the airwaves of Savannah. Striving to continue the goals of those who taught him how to do what it takes to get the job done.

KEN KNIGHT
Ken Knight has the POWER!

Before he received the opportunity to manage or buy his own station, Ken Knight fought the powers of management by making deals unheard of by the race before and taught his peers things they needed to know to make it in the industry. Ken Knight worked with his peers and believed in sharing his knowledge and abilities with anyone who needed his expertise in handling a situation. As he hypnotize the community of Jacksonville, Florida, by looking after them and seeing that things were done the way they should be, he was one of the few to be recognized by the community, both black and white during a time when this generally didn't happen. It was virtually unheard of to have a street named after a black person in the sixties, but; Jacksonville, Florida did and Ken Knight Drive is still on the city map today.

Ken Knight was a disc jockey who hit the airwaves of the south strikingly throughout the entire black community and became one of the South's most popular Disc Jockey's. Also known as one of the 'Bad Boys'. Working stations WERD—Atlanta, WROD—Daytona, and WHRC—Jacksonville was not his only challenge as he later became host of a local Sunday morning show in Jacksonville, Florida which aired local and regional gospel choir talents, ballet, and tap from all across the southern region. In 1962 he was highlighted on the cover of Open Mike Entertainment magazine as doing the following:

FLORIDA D.J. SCORES T.V. FIRST... (picture of story on page 129).

He was a very important factor in the community of Jacksonville, Florida. There are many who can still remember listening to his voice over the airwaves as he read letters to Santa Claus from the children of the city over the radio. Many can still remember the rush of happiness that filled their hearts, hearing their names called and his announcing their names as he described how they had to be good as they were waiting for Santa's presence once again.

In many homes it was a Sunday ritual waiting anxiously to see who Ken Knight was going to have on his weekly Sunday show. It was quite a joy to sit down in front of the television early on Sunday mornings and pick out people you knew from the community and also to look at those who were from other southern cities and compare them to the choirs in the hometown of Jacksonville, Florida. This was an important occasion in households as many got ready for Sunday School; and sometimes, would see some of the people who had aired for the show on their way to church and would smile and tell them that they saw them. This was a very exciting event for the community and for the ones who got the opportunity to showcase their talents.

Another joy bought to his nieces and nephews was the joy of getting free tickets to concerts. During this time the local disc jockeys would come out on the stage as the Master of Ceremonies and present the performers for the evening. And you could get a chance to see your very own Uncle at work, it was a thrill. Firm to his belief, the girls in the family were never allowed back stage to meet the stars, because Ken Knight thought that this was not the place for young girls. Because of this, it did arouse jealousy towards the boys from the girls in the family. But because the economy was a struggling one for the black community at this time there wasn't too much worry about this issue because they did get a chance to see the Jackson Five the first time they arrived in Jacksonville with the Commodores playing along with them, and they also got the opportunity to see the Temptations give a show which they would never forget. Also, from time to time he would get them tickets for a chance to see the Globe Trotters and the Ringling Brothers Circus. There were many other events and because of his kindness, the children, (who were always first in his book), got a chance to see and experience a lot and today, they can only hope he knew how happy his generosity is still remembered, they can only hope he knew how happy his generosity made them feel.

During this time blacks didn't have to worry about drugs or alcohol abuse as they do now, and it was fun for everyone to go out and enjoy an event without worrying about anything dangerous happening before or after the concert.

One of the younger sets of Disc Jockeys received the opportunity to work with Ken Knight and some of the pioneers of radio during early 1970. Bill Baker openly explains that the present day Disc Jockeys are now starting to use some of the older styles from the fifty's in order to survive and they see that the format works. Bill also explains that by the development of a company called 'Vee Jay', the earlier Disc Jockeys like Eddie O'Jay were able to play black music for the first time.

Bill went on to explain that working with stations such as WERD and pioneers such as Ken Knight has given him the experiences he has needed to stay in the market including winning the Arbitron Rating of Men and Women in 1972 on WORD radio in Jacksonville, Florida; which was once owned by the Cohen brothers who also owned WIGO in Atlanta.

During this time of the black Disc Jockey's career, they were beginning to face the problem of not getting paid enough. Bill recalls leaving the country and western station he was working for on a part time rate and went over to the black station for employment, but, before getting his position at the black station there was a problem concerning salaries which would have to be cleared up. The Program

Director explained that before anyone else could be added to the payroll they were planning on getting a salary increase or having a walk out. Being paid only $175 a week at this time for the black Disc Jockeys was an insult to what their white counterparts were making. Thus, the black stations were beginning to ask for more money from their white station owners. Bill explained that the station owners were called and informed that if they didn't get the requested pay scale in which they were asking for then they were not going to show up for the 6 a.m. shift and the station would be dead all day. Bill recalls that these bold Disc Jockeys did just that and the station was off the air for one day and back on the next day when the owners decided to give in.

Bill recalls working with such greats as Willie Martin and Joe Bailey at WPDQ while in Jacksonville, Florida and others including Johnny Shaw and especially Ken Knight who taught him how to handle people. Bill explained that if someone was trying to get black owned station's FCC licenses, that, Ken Knight would be the person he knew that they could turn to because he was an excellent planner, and he also commented about the amazement of how Ken Knight would operate issues and make things go his way. Lastly, Bill remembers Ken Knight's efforts when radio station WPDQ decided to go to the black market and hired him as a consultant. Bill remembers this all, and closes with saying . . . 'A Bad Boy!'"

Ken Knight was born in Headland, Alabama and became the first black radio announcer in the south at WROD Daytona Beach in 1947. He produced the popular Knight Train show on WRHC that some have said he had at least $150,000 worth of advertising followers for the station which later became known as WERD after the call letters were ceased to be used in Atlanta. He became vice president of WRHC before moving to station WPDQ where he became the General Manager. He founded the ARattler Network@, a 10 city radio network which would broadcast the annual Orange Blossom football classic from the Orange Bowl in Miami. He was also one of the founders of the National Association of Television and Radio Announcers.

But Ken Knight's adventures began far before the days of Jacksonville, we shall examine more as we discover that the first black owned radio station in the country was not in the north, but, in Atlanta, Georgia in 1949 where he worked as the Program Director.

THE FIRST BLACK OWNED RADIO STATION

The First Black Owned Radio Station
(Atlanta, Georgia Makes It Happen)

The South was rising, shining, and awakening all in one day as the city of Atlanta was recognized as having the First Black Owned Radio Station and let everyone know that this can be done!
This they did, and did it well.

The news heading tells it all . . .

***

Radio Revolt In The South

Negroes invade Dixie airways to help capture neglected billion $$ buyer's market. This may mean a new era in better racial understanding.

Photo by: Dick Saunders Text by: Major Robinson

Our World Publication—Source of information.

Without benefit of gunfire the South is witnessing a bloodless revolution that would cause the late Senator Bilbo to turn over in his grave. The factor responsible for this sudden outburst of democratic thinking was the almighty American dollar. Someone had to wake up to the fact that a ten billion dollar a year Negro buyers' market was being grossly neglected.

There was no such thing' as "segregated listeners' ears'" or Negro voices on the air, couldn't be jim-crowed. Jolted by the enthusiastic reception accorded the programs, this trend will receive impetus as more southern broadcasters aim for Negro audiences, that boost local sales.

Five years ago the situation shown at the top of this page just "couldn't happen in the south." Yet, it happened recently in Atlanta, Ga. And happening on a station owned by a Negro, it becomes all the more significant.

Many southern civic leaders are of the opinion that Negroes on the air have done more to better racial understanding in one year than hundreds of orators from "up north" in the last decade. With Negro radio audiences' tastes the same as whites, most of the programs are free of offensive commercial spiels such as skin whiteners, bleaches. Our World shows you the people responsible for this southern radio revolt. The news heading read...

Atlanta Negro makes history when he buys and operates own radio station.

It's safe to say that on Oct. 3, 1949, in Atlanta over 99 percent of Negroes owning radios were tuned in to station WERD. On that day the first and only-Negro-owned studios started operating. And even those who had been skeptical of J.B. Blayton, Sr., hastened to congratulate him for making history.

Quiet and reserved, Blayton, only Negro CPA in the state and prominent businessman, had been thinking of a station since he started the Midway Radio & Television School in 1947. "Where will the graduates work?", was the question that confronted him. And when a local white station's plan to build studios in the colored section failed, he was more determined.

In the modern equipped studios at 274 1/2 Auburn Ave., in the heart of Atlanta's little Harlem, WERD sends out their varied programs to an estimated 240,000 Negroes designed to fit their listening needs. And a recent survey by Atlanta University on radio habits revealed that WERD has a 40 percent white audience.

This convinced local merchants that time bought on WERD would mean merchandise sold off their counters. One large downtown department store even built modern rest-room quarters and lunchroom facilities for their increased trade from Negro women, who they discovered spent $120 million yearly.

Now, everyone was beginning to understand that no longer could we wait for stations like WOR to let us voice our opinions and play our music, we invested in ourselves and won!

During this time in Atlanta, the city had Pullman cars, and some of the streets as we now know them had other names, such as, Bird Street was renamed to Hill Street. And today, we have the many malls; but, in 1949 all the stores were on Alabama Street. The careers of most of the race during this time were the jobs of Pullman Porters, Cooks, Postal Carriers, etc. (whatever could bring you the best money).

Instead of driving all over the metropolitan area for various clubs, many of the black middle class would find all of their pleasure at the Arts Fellow Building, across from the Herndon Building. Many of the sororities and fraternities found great pleasure in going to the Arts Fellow Building, which, was designed with a roof garden, and inside the dance room, there weren't tables and chairs, but, benches along the walls, because all of the space was used for dancing. Surrounding the Sweet Auburn area as passerby met some of the town's Disc Jockeys as they attended and hosted some of the affairs at the Arts Fellow Building, would be a surrounding neighborhood of all necessities, Beauty Shops, Doctor's Offices, Chinese Restaurant, and several other business offices which were supported by the community.

And the influence of the music played from WERD influenced, The Top Hat, later to be named, the Royal Peacock. Atlanta was taking the air waves by storm in the south and the community represented by Lawyers, Doctors, Teachers, and Preachers listened, supported, and responded and with well support.

WERD is truly remembered for all of the wonderful types of information inwhich they were broadcasting to the community and even more by one of the methods they used in order to get the white community even more interested in what they were doing. The key this was to have the best jazz broadcast in the area and this they did.

WILLIS SCRUGGS

As the evening jazz show was broadcasted by the now retired teacher, Willis Scruggs, members of the community listened amorously as they walked the streets of Atlanta and as they arrived home for dinner. Willis Scrugg's jazz show would start as soon as he would leave the school where he was employed as an instructor and hastily he would arrive at WERD to play selections from his own collection.

Willis Scruggs was very proud of having this position because it allowed him the ability to play the type of music that he also performed as a jazz musician. Sometimes, Willis would broadcast until 7:00 before getting home from teaching all day and working at the station, and sometimes he would get home earlier. During this time the sunset was the designator for telling the AM channels that it was sign off time. As the sun begin to slowly set, Willis would play the Ink Spots, or Count Basie, or maybe Joe Williams and close out with a little Dinah Washington and when these particular stars were in town they would always come by and meet the Disc Jockeys at the station.

Willis Scruggs remembers when the kids would get out of school and call the station and make request of their favorite songs and sometimes they would just want someone to talk to. As WERD was heard all over the community Willis Scruggs would use conversation to comfort some of the young men when they would call in and tell him that they wanted to hear something that could relate to their girlfriend leaving them for someone else. Willis would then tell them stories about how his baby left him, even though, she hadn't and would slowly begin to introduce songs from his private collection, because singers like Bessie Smith didn't have LP's and 78's were a little too scratchy for his liking. Willis would soothe the broken hearted with songs like, "Big Mama Thorton's", `My Baby Left Me . . . Standing In The Back Door Crying . . . " or sometimes, the Ink Spots would come by or even one of his favorite cousin's—Nipsey Russel would come by and talk to the audience. Willis enjoyed this very much and for three years this was a part of his life . . . Teaching, and being a Disc Jockey during the jazz hour at the first black owned radio station.

Willis Scruggs recalls how the influence of jazz was very big for the white community and when performers would come to town and he would help with promoting the artist, there would sometimes be a section reserved for the whites and sometimes the whites would have one night and the blacks would have another night to come see their favorite artists perform. Because WERD started out establishing itself as a classy station and not trashy it was supported for it's outstanding way of reaching it's public and when the public wanted to meet them at the particular dances, they would be there to say hello, if they weren't working the show that night. Atlanta had a lot to be proud of and so did the community as they listened to the staff of Disc Jockeys that woke up the air waves in the city and changed radio history with their creative ways of doing things the best they could and never did they let their listeners down. The people knew this and remained loyal until the station's letters were finally put to rest.

There were Disc Jockeys and then, there were Disc Jockeys, Disc Jockeys sold the stations, not only with selling the commercials, but, also with their creativity and voices and there are still many living who can remember their swaying voices, such as, Jockey Jack, later to be renamed to 'Jack The Rapper'.

JOCKEY JACK aka "JACK THE RAPPER"

Jockey Jack or Jack 'The Rapper' Gibson
Another one of those Bad Boys

A feeling of personality permeates the room when one would interview Jack "The Rapper" Gibson. Before reading the facts to Jack's story of his days in broadcasting, here are some of his career highlights.

Jack Gibson
Born: May 13, 1920, Chicago, Illinois
Pioneer black disc jockey, writer, public relations manager, publisher of Jack the Rapper, promoter of Jack the Rapper Family Affair, the largest national black trade conference. His career began as a radio

disc jockey at WJJD in Chicago, one of the original 13 black DJ's in the country. In 1949 he became a DJ at WERD in Atlanta, the first black owned and operated radio station. Gibson held DJ jobs at several other stations around the country before being named national director of promotions for Motown in 1962, where he remained until 1968. In 1968 he joined Stax in public relations. Jack's roots are from the same school as Joe Medlin and Dave Clark. He ranks high in the music industry with a multitude of followers and success stories. In 1985, Jack was successful in leading over 230 radio stations to honor Stevie Wonder on his birthday. Jack Gibson challenged radio stations as a result of the government of South Africa banning all Stevie Wonder's records because he accepted his Academy Award in the name of Nelson Mandela of South Africa. Jack Gibson is a strong supporter of the National Black Programmers Coalition; vice president and board member of the Black Music Association; member of the Blues Foundation; and advisor for Who's Who In Black Music. He is in high demand for speaking engagements at colleges, universities, and before organizations. Jack received a Bachelor's Degree in science from Lincoln University in Jefferson City, Missouri.

After flying across the country to attend Whitney Houston's marriage to Bobby Brown, Jack rushed back home to finalize his 1992 plans for the Jack The Rapper Family Affair Convention. Jack admits that he always finds himself coming back to Atlanta for the annual convention because, this is where it all started to happen for him and it's still where things are happening for him. (Unfortunately, because of problems with the 1993 convention, the 1994 location was changed to the city of his home office, Orlando, Florida). Jack admits that he was one of the first blacks to get started in radio (personality radio that is), and with open honesty he explains what people expect pioneers to look like—"straggly bearded". Jack explains:

"1st of all we weren't black we were negroes, and sometimes titled colored. "During the early days of my career I wasn't into broadcasting,

I started out as an actor. I spent some summers working with the summer theater. I remember being told that I had the talent to `shred the boards' as the term for the stage was used then. And a very famous actress I appeared with told me that I wasn't going to make it because I had something against me in making it in Broadway and Hollywood, (At this time I had only done juvenile roles and characters)."

"I recall her asking me to her dressing room and told me, `I don't want to burst your bubble but you have two strikes against you, number one, your color is going to keep you from making it, you're not white enough to be white and not dark enough to be a negro, so you're out'. She went on to tell me not to give up, and also told me that radio is a very strong thing. I remember her telling me that they had drama in radio and she had some friends who could hook you me up with some people in Chicago."

"Chicago at the time was the mecca of all radio drama. Deluxe Radio Theater, The Shadow, Joyce Jenkins, all of those different kinds of shows, so she hooked me up with somebody. I got to be on radio soap operas as a character actor with soaps everybody was listening to at the time, such as, Joyce Jenkins—Girl Intern, Here Comes Tomorrow, Life is Beautiful, Bob Perkins—Young Widow Brown, which all of them were 15 mins. in length and you could do four of them in a hour. This was the beginning of what we called soap operas because soap companies sponsored them. I used to do shows everyday. One day in the green room we were kidding and I did a take off of something, I don't remember exactly what and the Manager came in and said who just did that take off on such a such a thing? I said I did. I thought they were going to fire me. He said no we're not going to fire you, I got a man here who wants to know if you know anything about music. I said yeah, I led the band in college. I was sort of a drum major. I would shake my hair and it would fall down in front of my face ala Cab Calloway. (Yes, I led the band in college but I didn't know the first thing about music, I didn't even know how to keep time). I talked about my liking the style of Jimmy Lunsford, Duke Ellington, etc., he said how would you like to do a radio show and play music and make $65 a week? I said huh! when do I start? (At this time they didn't call us Disc Jockeys)."

"This was almost at the very beginning of them calling us Disc Jockeys in 1945 or 46. I began airing a show from 11 to 12. The show after mine was the guy that founded the Today Show, the first Today show host, Dave Galloway, in fact he tried to get me to come to N.Y. with him when they were taking auditions and the old not being black enough thought in my mind held me back. And from this point on was when I got involved with radio."

"As time went on and my career was beginning to take off, & I decided to marry. The best man at my wedding father was a CPA in Atlanta and he had just bought a radio station which was bankrupt and was sold on the steps of the Fulton County Courthouse, he called his son at the University of Chicago and said I want you to come here and operate the station. He went on to say, you got a friend Jack that's in radio, and he said well, Jack's an actor. He told him to bring him on to Atlanta with him, and that's when we came to Atlanta in 1949, and that's when they bought Ken Knight up from Daytona Beach to be the first Program Director of the first black owned radio station in the country, this was also when they began to call me `Jockey Jack', they took off the word Disc."

"About 16 years ago, (the 1976 time frame), I decided to start my newsletter magazine. I wanted to have a hook name in the writing thing as I did in the talking thing, as I call it. So I thought . . . I talked, and I said to myself that's rapping. Jack the Rapper—smooth talking to the ladies, so I said that was rapping. I said to myself, `Jack the Rapper', that sounds good. Of course the new guys in the industry today think that I'm an old time rapper and call me the grandfather of rappers since I'm in the twilight years of my life. But they must know that there were 13 of us (bad boys), (known as the record breakers) and only 2 of us are still living, Hal Jackson who's Master of Ceremonies for Miss. Black America and myself. Everybody else has gone. As for the others on the air earlier they haven't come forward or possibly can't—today."

"My career has given me more than I ever expected; longevity, beautiful friends, family, being in demand for things, being asked to host different events. I was given the opportunity to speak at Mary Wells funeral, which was highlighted in Jet Magazine, and even though this was a good bye to a dear friend, I made the funeral a celebration instead of a funeral, which was right before the Reverend did the prayer. I talked about our days in Motown, and you could tell that everybody appreciated it, and her family thanked me for lifting their spirits at such a mournful time."

"I not only still get the opportunity to work with those who came up the ranks with me, but, also with the young set that's out there today. I feel that my career has allowed me to still be involved with the younger people and to be involved. Sometimes when you feel there's no hope, some old voice can come along and kind of tell you some things that might bring new hope."

"During my days of early broadcasting we were able to make or break an artist. I was successful in assisting with getting, `Lonely Tear Drops by Jackie Wilson, and You'll Never Walk Alone by Roy Hamilton, known by air play over and over again."

"In my early days of broadcasting, there was no competition among the Disc Jockeys. In later years, for some, but by that time the others along with me, had arrived at such a level that you were automatically a super star and so you shone brighter than anybody else in the town. The Disc Jockey's of the cities became the spoke person of the cities. People listened to you, for me it was talking to my listeners at Buttermilk

Bottom in Atlanta. We would visit the neighborhood and talk to the people".

"This was a time when the Disc Jockey was the spokesman, the leader in town, we didn't have a Martin Luther King, Whitney Young, Jessie Jackson, or Ben Hooks, to talk to the community yet, the only leader we had to look up to was the heavyweight champion of the world, and that was Joe Lewis. That was the only one we could look up to, so therefore the disc jockey was the leader of the town, and if something went wrong or somebody couldn't pay their rent, the Disc Jockey would hold something on the air and get money together and tell the world across the air waves, we paid Ms. Jone's rent. Whatever city the name was, our powerful voices became the boss for the community and the Disc Jockey could command everything, right down to getting more pay and this is the way it was for the beginning of our careers, as Personality Jocks. If Disc Jockeys wanted more pay and the owner acknowledged, but not without telling them what records to play then this would be the end of personality radio and as the history of radio tells you, this is what happened, the mighty dollar won out, we no longer had control over the community as we had before and a new breed of broadcasting was developed."

"Even though Personality Radio was taken from the air waves, it is still remembered by me and others who really care about what it all meant. I can never leave out such an important part of our history and not give it the recognition which it should have been given, therefore, I award at each one of my annual conventions an award to let people know who was really here before them, making it all possible. Each year I name an award after my brothers and peers who have passed on that I was very close to. I saw this as a way of letting people know who they were. We do it in 10 year segments and I'm on my 2nd set now, including the ones that followed them. Ken Knight was on 10 of the awards. They're part of our black history, and are important names to be remembered. I named the awards not out of selfishness, but, because I knew all of the people. It was who I thought personally who should get this award. I handled the first fifteen years. After recognition was given to all of the personality Disc Jockeys I began to name others in the entertainment industry, such as, `The Florence Ballard Award' which was for her to be remembered as a Supreme not as a heartache or anything else, so that other female singers who get this award would have the hope of reaching the same talented goal."

"In my earlier years, I wore a jockey suit to promote myself. I was in Louisville, Kentucky which was where they had the Kentucky Derby, so I had some silks made. An Italian guy at the time made silks for the Horse Jockeys, and he designed me a set of silks. Irgles 92 Beer, who was one of our promoters had the color of red and black and I wore a red and black jockey suit with the pom pom and hat.

I did a man on the street program 15 minutes everyday in front of the black theater and would pass out free beer. At that time you could give out free beer. Once I took a picture in my outfit on top of the turntable and used the microphone as the horse, and this was my job, it was all in fun, and believe me the promotions worked."

"Radio was all people had at this time and Personality Radio is what we used to get their attention? Personality radio, was more than talking to the community, it was also giving the community originality over the air waves by never introducing the record the same way, being original. From the program book we would look at the next commercial ad and act it out, such as Sweet Pea Snuff, and Bull of the Woods Chewing Tobacco. I had a cow and it would moo and I would say, now if only I could have a taste of Bull of the Wood chewing Tobacco, I'd feel so good and I'd turn it over and it would go moo and someone would say, who ever heard of a bull mooing? Well, we couldn't fool them all of the time. We still had fun as we ad libbed all of our commercials."

"People's imagination of you as they listened would be totally different when they met you. When they would see you in person after listening to you on the air they tell would tell you how much they love hearing you talk that trash. One guy in Lithonia, Georgia at the black country club they had then met me through his wife and told me that I was the first yellow N. . . . they ever let come to Lithonia. When they saw me they couldn't believe I was the same person who talked so much crap. This was the job of the Personality Disc Jockey, to serenade our audiences and sometimes influenced housewives from doing their house work and listen to . . . Blue Velvet instead."

"Our sense of community was beginning to spring up everywhere and because of this we were able to get help from stars when we needed it. Once, James Brown and Ike and Tina Turner performed for us at the Regal Theater in Cincinnati. We had booked Ike and Tina and she wasn't able to perform . . . . James Brown found out about our booking problem and came over and played for us so the public could get their monies worth, we got James for nothing and you can't do things like that today. All of the Motown acts would play for free, in Cincinnati. We had 4 shows titled, `The Midnight Ramble' and charged $2.00 for the show. Berry Gordy would send his acts to us to perform, from Stevie Wonder to the Supremes. This gave the Disc Jockeys a chance to be in promotion and control the radio with whomever the company would send to us for free, and this was their way of breaking a record. We in turn would clown with the acts and entertain the audiences as we promoted the acts. Also, they would soon get air play."

"Radio works! It's powerful entertainment and the way you use your voice creates an atmosphere of comfort. One of my shows was, `The Sound', which was jazz, it's like the pre-quiet storms which are played now."

"Personality radio worked, when I would tell all my women listeners to put on red dresses and look pretty for Jockey Jack, all of the street would be filled with red dresses. We were very community oriented, and this type of creativity would also tell us how many listeners we had when we would see for example, all of those red dresses. The older generation remembers this type of broadcasting, we were part of their house, we laughed with them and cried with them. We were a part of each other's life. We had a lot of power, we weren't told what to do, we played the records and went by the station format we put together, we would read the Farmers Almanac for those who had interest and tried our very best to satisfy all of their needs."

"As for radio today, it is really not into the community as it was in the past. Back then everybody knew each other and everybody helped you out who could. Things of this nature was how we helped out performers such as, Gladys Knight and got her started, and she never forgot this. When you take the Personality away from radio you make it more like juke boxes, and of course the demand for more money would be the next issue."

"Of course anytime you do anything in life there are skepticals, and I'm sure there were some when WERD came into play. But we had a plan to make sure everything we did worked. WSB was already around then and everything they did on one day, we did the next day, and the next day it was like listening to the black version of WSB. We became very involved in everything we did. We gave the community everything and we imitated them, and it worked. 90 % of the blacks didn't care what WSB was doing, they waited to see what the black station would say about situations and whatever we said was the word. I had a sound effect record with a ticker tape on it and I would say it's high noon and the news is being reported by Jack Gibson. I would say something like oh you don't want to hear about that, because that didn't happen but let me tell you about Hunter street . . . and in between this we would input the ticker tape recording. We didn't stop here, we went on to classical musicals. A colored station playing symphonies? Mozart, etc. It was crazy. We never thought of it as anything much then, but we look back on it now and see it as one hell of an era."

"All of the prisoners would write to us and we would talk to their families over the air for them. We would call it the vacation spot at the end of McDoughn Blvd. and all the guys that got out of the State Penitentiary would call and tell everyone we're on the way back to Detroit, etc., yes, this was a time when we reached out to all."

"As far as retirement from my publishing and The Rapper Convention, I don't plan on retiring. I have laid out good ground work, and my convention was a convention of my thought, to help the young to have a place to network, talk and swap ideas. The energy that I have given it through my personal self I hope that it will continue to grow bigger and bigger and 10 to 20 thousand will come to discuss our business. My legacy would be . . . to let it grow and to utilize the energy that I have given. I know when they portrayed the first thirteen, aka, The Bad Boys at the Smithsonian Institute, they put us in as a relic, as the old dinosaurs, so young people could see the relic."

In 1996 the convention was 20 years old. Jack felt strongly that he fathered the business, and radio helped him to stay young. "God has given me a task to do—look after his children and that's what I'll do until I'm no more."

SHELLY STEWART

While Jack Gibson and Ken Knight were busy making things work for the city of Atlanta, Shelley Stewart was starting to make it all possible in Birmingham. Shelly Stewart started broadcasting in August of 1949 at radio station WEDR in Birmingham, Alabama. Station WEDR was named after Ed Reynolds (the owner), such as many of the other stations during the 1949 era which were named after the white owners, such as, WJLD which was named after J.L. Doss. In the early 60's this method of naming was taken away so that more easier remembered call letters could be used.

WEDR was the first black programmed radio station. Which had trouble in it's beginning as the KKK tried to keep them off the air by tearing down the station's tower. And as a result the station needed the support of the local officials whereas they had not been receiving the support which they needed but, he and others kept the pressure on. WEDR gave him the opportunity to speak to the black community. During this time there were some who felt, (as mentioned in other interviews) that blacks had to sound just as the whites on the air when they did their daily broadcast. At this time was when the quiet low voice began to become a part of the speaking stream, with voices of those such as, Paul E.X. Brown, Eddie Casslebury, and Walter Eglin Jr. during the 1949, and 50 time period. As a result of this the station's mail was showing the popularity and as the mail was read over the air, Shelly Steward knew that this was ok, but, he also knew that there still was something missing, therefore, he began to use the louder voice of broadcasting, and began to reach out to all of the community—prostitutes, those living in the projects, everybody was included, because he saw that they were part of the community and wanted to reach out and touch their lives as well, thereby, he began to use slangs, such as 'great goobly-woobly', and so fourth. Shelly was labeled as someone who wanted to play and play and play and therefore received the name of "Shelly The Playboy".

Shelly (The Playboy), vivaciously recalls the white fan club that supported him from the white community and named it the 'Shelly The Playboy Fan Club'. Shelly enjoyed this because of his belief of being in an industry entitled 'Mass Communications' and not 'Class Communications'. By having this thought Shelly treated them like they were people because they are. 'I never believed in looking down on other people, and I believe that all people are equal and should be treated as such."

Shelly realized early in his career that people such as the prostitutes, poor people, people employed in community service positions such as policemen and firemen were human, and needed to be treated as such and therefore he would talk to them and others as well. And by his being a personality disc jockey, he knew that this was the right thing to do. "Talk to the people and let them know that someone understood and someone cared about their issues and concerns." Shelly recalls telling the neighborhood, "Goooood Morning . . . " and as a result this trend began to spread to other stations across the country. (Bringing out the personality of his trade was what he loved doing the most).

During this time in the South some stations were a part of a radio chain named the 'OK' group owned by Stanley Ray who was a white male that took ownership of stations WAOK, and WLOK in Memphis, WVOK in Baton Rouge and others stations which were to become a part of the OK chain. And as a result of this station name pattern being developed there ended up being a lot of disc jockeys which would have the same air name. Some of them were called, for instance, "OKY-DOO-KEE" or "Diggy Do", and the persons who had these names while broadcasting for the OK group would leave the name with the station when they seeked employment somewhere else, then someone else got your name and thereby, would be introduced as "OKY-DOO-KEE", because the station would not let you use a name which you or the community thought of. Sometimes, Stanley Ray would come and listen to Shelly and would go back to his stations and tell them you got to sound like this guy, but don't talk about education, don't talk about politics, don't get involved in the community, (anything controversial), but, Shelly did all of this and survived and was supported strongly by the community. As a result of always being a loner, Shelly is one who believes in doing what's necessary to get the job done and by doing this received great applause from Birmingham, Alabama and as he went over to Jackson, Mississippi and tried his talent there, he became an immediate success, and took the city of Jackson, Mississippi by storm for six months, and later received his first actual contract after he opened in 1959 in Columbus, Georgia radio station WOKS for six months, and then returned to Birmingham, Alabama at radio station WENN, which was a black station that opened in 1950 after WEDR. Later WBCO opened in Bessimer, Alabama, twelve miles from downtown Birmingham started out with such great jocks, such as, Larry McKinley, and Bruce Paine Jr. from Jackson, Mississippi who started with WBCO when it opened and later changed it's call letters to WENN.

Later in 1966, Shelly took a one year contract with WAOK on Edgewood Avenue in Atlanta, Georgia, and afterwards returned to Birmingham. As a result of all of the training he received, Shelly used what he learned so that he could finally become an owner of himself, and as of today still owns WATV, and today he has a syndicated talk tv show in Birmingham.

Even though Shelly had the most highest ratings, and popularity in the area, he admits that his color stopped him from receiving the money for the spots and explains that he never could get the recognition which he deserved from the media. And it was people like Ken Knight who would support him by telling him that he had more guts than anyone else he had ever met in Birmingham, Alabama and encouraged him to stay there and do the things which were necessary for him to do, so that he could make things happen for the community. By hearing comments such as this, Shelly kept his vision on track, and kept his strong backbone as he recalls other well known disc jockeys; such as, Martha Jean Steinberg, and disc jockeys such as Malcolm Moore, and Butterball, and recalls the power of personality they used as they broadcasted from their radio stations and began to permeate the south in stations of cities such as: Birmingham, Alabama;

Memphis, Tennessee; Baton Rouge, Louisiana; New Orleans, Louisiana; Jacksonville, Florida; and Miami, Florida.

Shelly still remembers the Magnificent Montaque who he gave the phrase 'Burn Baby Burn' to, and after Magnificent Montaque left Mississippi and started in New York, he looked up Shelly Stewart and asked for his support and help. Shelly remembers sending him tapes and tells how 'Burn Baby Burn' actually started. When Shelly Stewart would broadcast at his station in the south, he would love to use the phrase 'Burn Baby!' and that's the way I would introduce some of my songs. We were talking about the song being hot and whites thought he was saying that he wanted the city to burn during the riots.

The disappointment of some of the radio stations today from Shelly's point of view is because the news, talk shows, and community concerns have been taken away in the radio broadcasting industry and today admits that 'boogie' is all we hear and it doesn't necessarily mean that it's what the community wants.

Shelly still honors WERD as the first black owned radio station in America and WEDR as the first all black programmed radio station in America. And today he wants to reiterate this fact and thereby should be recorded as such.

And to finally honor his years of dedication and commitment to personality during the month of August in 1993, Shelly and peers were inducted into the Black Radio Hall of Fame at the annual Jack The Rapper Convention. Eddie Castlebury, Hal Jackson, Martha Jean Steinberg, Shelly Stewart, & Charles Scruggs.

E. RODNEY JONES
Earl Rodney Jones

E. Rodney Jones has held the titles of Program Director, air personality, and is noted as a legendary broadcaster. This legendary radio personality could be found at WTKL in Baton Rouge, Louisiana. Along with holding other successful positions and employment from Chicago to New Orleans. E. Rodney Jones was the most popular disc jockey in the Chicago market for several years. E. Rodney Jones worked as program director of WYLD-AM in New Orleans. In 1984 he got together with Buddy King at KAAY in Little Rock in order to produce a national Blues syndication program called Blues Alley. The golden voice of E. Rodney Jones has been emulated by many in the industry and he still holds his record of personality, and charm.

Working in the field of communications via the umbrella of radio broadcasting has been a memorious statue for E. Rodney Jones to behold. "By being able to communicate to those who are listening to you and relying on you as much as they would 911 with situations such as looking for a lost child and other emergencies . . . " was quoted as one of the exciting parts of his career as his unselfish dedication was acknowledged for three years as he was awarded program director of the year with Billboard magazine, and also highlighted on the cover of Black Radio Exclusive magazine. And not to exclude his position of Master of Ceremonies for the Black Hall of Fame of radio announcers.

As E. Rodney Jones developed his career in Chicago there was a complete change in radio as management intercepted his comfortable broadcasting position and asked him to go to Louisiana for six months as he would soon find that his six month transient position would turn into eighteen months and eventually veer him into the world of record promotions and back into radio again.

During this time he would find himself going from one station to another while working the cities of New Orleans to Baton Rouge, Louisiana. As he began to adjust to this new venue of his career, he once again experienced the selling of another station and moved over to Little Rock, Arkansas through the assistance of George Frazier. After eight months, the station at Little Rock, switched their programming and without letting this be a low point in his career he took the opportunity to start his own record label Midtown Promotions whereas he moved back to the state of Louisiana. As a result of this move things were beginning to work well for him and as of three years ago on April 1st he has enjoyed being in the top ratings in the city of Baton Rouge, Louisiana and still can be heard today.

Even though E. Rodney Jones has found himself moving from state to state throughout his career, he hasn't taken this as a low but will admit one of the lows which have been addressed by most of the pioneer disc jockeys and that's the removal of what could have been one of the most strongest organized calibers for disc jockeys which was the ending of AFTRA and NATRA as they were removed from lack of support as being part of the industry. As a result of the removal of organizations such as these, changes in station's formats would put you out of work and this has effected he and his other peers positions as disc jockeys. One example of this occurring in his career was when he and everyone working under him were released from their positions at WVON after serving twenty years of dedication to the station and as the station was sold to the Gannet organization they bought in their own people and everyone under him became unemployed. But through it all E. Rodney Jones still has special feelings about his career as he recalls working with Hitsville, USA as they spent many tiring hours trying to develop something that they all had at heart and that they all wanted to work and provide work for others. Shortly after the hit song, 'Shop Around', by Smokey Robinson and the Miracles became an instant hit and the rest of this we all know as the Motown story.

Another memory was when the Regal Theater had the Motown Review and acts such as a young Stevie Wonder, (thirteen years old at the time) performed and as time passed E. Rodney Jones recalls how this person who was so young in his career remembered him later in his adult years, as Stevie Wonder once called him and sang a special tribute to him during an award presentation that was given to him, and as Stevie made him smile throughout the conversation he began to recall their initial meeting from when he was thirteen years of age and has found that it is through the embellishment of honors such as these which makes doing what he likes the most a memorable event.

Even though there were fun times such as mentioned above there were definitely responsibilities. Disc Jockeys in the earlier years had to broadcast around the Unlimited Broadcasting format, which meant for example, if a baseball game was on the air at three in the afternoon, then you would broadcast up until three, if the game didn't air then you didn't go on the air.

E. Rodney Jones recalls when another disc jockey would recreate a previously played baseball game in order to get on the air as he would use AP wires, ticker tapes as he would read the script from the game and pretend to be giving the game as if it was actually being called play by play.

Even with this responsibility they had to tie in another, which was the responsibility of advertising. As far as most of the advertisers he portrayed over the airwaves, he recalls how they would be given a matchbook to do a sixty second commercial and when he would get ready to start the commercial there would sometimes be an advertisement without the address and he and other disc jockeys would have to rush and find the address in the phone book so that they could announce it for the commercial spot. For many years he did the Schlitz Beer and Kraft Bar B.Q Sauce commercials.

This responsibility was also a part of what Personality Radio was all about. Personality radio was what the community expected from he and his peers of radio and through this powerful entity he made the community remember what he was there for—them! "The Air Personality communicates with his audience", is what Personality Radio means to he and his other black pioneer peers of radio admits E. Rodney Jones.

And as he tells those seeking employment in this remarkable area of communications today he advises them to get a high school diploma, college degree and then a masters degree in order to make themselves more marketable in the industry and will keep their individual status strong in the industry.

E. Rodney Jones believes that his strength of maintaining a position in the industry began with his broadcasting roots. As he started in radio 42 years ago in his home town of Texarkana, Texas and then on to Kansas City where he worked as the only black hired at radio station KUDL and almost immediately he became a very dominant figure in the Kansas City market. As he developed his career he later became master of ceremony at a place called the Orchid Room. While taking advantage of what the reputation of Kansas City was known as at the time as the 'Jazz mecca of the world' and also called by some as the 'good time city', with the likes of Charlie Parker who was a premier performer of his hometown and also Joe Thomas and his band he strengthened his connections and developed his career more and more as he began to introduce many of the well known greats to the community. These and other local greats such as Andy Kirk, Joe Turner, Julia Lee, and Lorenzo Fuller were known as top performers. E. Rodney Jones experienced the highlight of promotional epitome when he received the opportunity to introduce the talents of such greats as Dinah Washington, Jimmy Lunsford, Count Basie and The Kansas City Five, Duke Ellington, Jay McShane and Charlie Parker. This was where you heard the great blues bands and where many careers gained their strength in the industry. Later his career was developed even more as he was able to work with his dearly departed friend Marvin Gaye, Diana Ross, Lou Rawls, Nancy Wilson, Barry White, and Al Green. And he still recalls how Al Green used to do a lip sync for him in Chicago as he was able to assist him in launching his career as well as many others. E. Rodney Jones vividly recalls Don Cornileius as a young insurance salesman in Chicago and as Roy Wood his news director gave Don the position of roving reporter he would later obtain the position of a disc jockey, as Don later made his next move to pilot his syndicated TV show E. Rodney Jones became the voice and the announcer for Don Crnelius but as Don moved to California he remained in Chicago and his neighbor who also happened to be a disc jockey named Syd McCoy followed Don and became the voice of Soul Train for Don Cornelius. This particular turn of events not only developed E. Rodney Jones but Don Cornelius and Syd McCoy.

Another strengthening factor for E. Rodney Jones is the format he has been able to maintain which consists of Jazz and Blues. As he strongly believes in preserving the heritage of African American Music, (Gospel, Jazz and Blues). And today he denotes that it is important that our generation of blacks know exactly where their heritage derived from. And for the past forty years or more of his career he has sent the message that the heritage of the black culture should be there for generations to come as they learn, grow and develop from it. Along with this he believes that we must keep what belongs to us and pass it on to our future generations with pride and honor.

With the honor of being a Member of AFTRA and later NATRA, E. Rodney Jones still believes that it's still not too late to pool our knowledge as a people and make things happen for our race.

Of course as mentioned before there had to be steps taken to develop and maintain your position in radio, but, there's one step that E. Rodney Jones didn't have to take because God had already given him this, the talent of his voice. Known as, 'The man with the Bedroom Voice', to some and to others 'The Golden Voice', he recognizes that God has blessed him with this voice, a microphone and the mechanics to project what the community has enjoyed hearing him broadcast. And as his 'Golden Voice' Penetrated the airwaves of audiences across the country for the past 42 years and still does today he closes out each day with the following quote:

> "I Believe in Life because with the absence of life there would be the absence of God. I believe in God because without God there would be the absence of Love. I believe in Love because without Love there would be the absence of you—the most wonderful audience in the world. And to you, I believe in you, because without you there would be the absence of me". and as he pauses he speaks once more by closing with, "Ladies, Roses On Your Pillow".

And the stations of KUDL-Kansas City, KXLW-St.Louis, WVON-Chicago, and WXOK (AM)-Baton Rouge have held the honor of airing such a wonderful voice to their listeners.

Another black who did not let his color stop him and proved that he could become what he had always wanted to become was that of Al Benson.

CHAPTER SIX

AL BENSON

*Al Benson
Mississippi Style*

One native Mississippian didn't let the fact of his color stop him from becoming a broadcaster, let's find out why not.

Al Benson was a native of Mississippi and developed an announcing style and programmed music which appealed to southern Black migrants. Benson acquired radio time through the brokerage system. He was heard throughout Chicago's Black community, first broadcasting from a storefront church; then jazz in 1945; jazz, blues, and rhythm and blues in 1946; and only rhythm and blues by 1950. His popularity among the masses grew so fast that by 1948 he was broadcasting rhythm and blues shows over four of Chicago's stations for a total of ten hours daily.

Chicago's Black community identified with Benson because he gave visibility and validity to the culture and music of southern Blacks.

One of his listeners was quoted as saying, "He was one of us. You could tell by the way he talked. None of that `uppity talk'. I guess that's why I liked him. And then, he played some good music." Benson's programming and announcing style had a major impact on Black and White announcers that became prominent during the 1940's and 1950's.

We shall learn more of Al Benson's charm and personality used in his career, as we read later on the career of Lucky Cordell who worked with him for many years.

But, first let's examine the life of Frankie Halfacre, a war veteran.

FRANKIE HALFACRE

Frankie Halfacre
Mr. Lucky

Because of an injury during the Korean War, Frankie Halfacre was advised to become a Disc Jockey upon leaving the armed services, and that he did, and is still doing today.

Frankie Halfacre's love for radio goes back when he was a patient in Valley Forge Army Hospital during the Korean War. Frankie was influenced by a disc jockey at the hospital by the name of Johnny Patterson. Johnny used to play King Pleasure's, "Moods' Mood for Love". When Frankie left the hospital doctors thought it would be good therapy for him to be on radio so that he would use the muscles in his face. Needless to say, he loved it.

When Frankie got out of the service he tried for 13 years to get into the business. Frankie continued to stay in the entertainment field by writing and emceeing, (especially at youth events). One special evening he was the emcee at a concert in the park and a white gentleman walked up to him and asked him if he was on radio. Frankie told him no . . . the reply was, "you have a job now".

This new job for Frankie was sixty miles each way every day; but, Frankie enjoyed the job at WWOW/WFIZ-FM and continue to take the long journey to and from work each day to Conneaut, Ohio.

Eventually, Frankie wanted to get on the airwaves in his hometown market and in order to do this he would have to buy his own time to play black music.

Frankie's show, "Lucky's Soul Kitchen" began as a one hour a week show and almost instantly changed over to a four and a half hours a day show.

Frankie used the name, "Mr. Lucky" because a young man by the name of Lucius Barnett starting calling him that when he worked at a playground and would always find four leaf clovers, and the name stuck with him.

Frankie was also "lucky" to have had the pleasure of coaching a track team and experienced the pleasure of coaching Terry Taylor of the Cleveland Browns and Laurie Gomez, the All American Distant Runner of North Carolina University, and they named him, "Mr. 1/2". "Lucky" or "Mr. 1/2".

Frankie Halfacre has given a lot of his life being dedicated to Black Music and putting it out front where it belongs.

One important event concerning Frankie Halfacre and his career is the fact that he quit working for one of the stations when they refused to let him play James Brown hit song at the time, entitled: 'Say It Loud, I'm Black and I'm Proud', by James Brown. And to this day, he doesn't regret any of it. Jet Magazine was there to cover the story and the rest is history . . .

JOHN RICHBOUGH

With everyone else before him,
What makes him such a legend?

Even though he was white, John Richbough is a noted legend for the time period of 1942 to 1973, he had become one of the most legendary R&B Disc Jockey's out of Nashville, Tennessee. (John's black peers in radio also requested his status in this story). John Richbough also known as "John R" was known at station WLAC as the Big Daddy of R&B Disc Jockeys. He spent most of his time promoting products such as; Hoyt Sullivan Hair Products, White Rose Petroleum Jelly, and Royal Crown Hair Dressing. Another one of his promotions was that of commercials for Baby Chick, and Ernie's Record Mart (For those of you who may not have known, this was a highlight during this time because he bought the accounts to the station and then he creatively turned around and produced them.

Not only was John Richbough the Big Daddy of R&B Disc Jockey's, but, he was also called, the 'hitmaker' as he launched the careers of Bo Diddley, Chuck Berry, James Brown, Aretha Franklin, B.B. King, Gladys Knight, Otis Redding, Wilson Pickett, and Jimmy Reed on his radio show. As we can see there are a lot of people who can thank John Richbough, 'The Hitmaker' for his dedication to the industry. Even today, being a 'Hitmaker' is quite an accomplishment from the recording side of the industry.

And now that we can see what accomplishments were made by E. Rodney Jones and John Richbough, we can take advantage of looking into the life of Alley Pat, another Atlanta Disc Jockey.

ALLEY PAT

What makes Alley Pat so special?
(Always Atlanta Bound)
Alley Pat

Coming from Montezuma, Georgia and moving to Atlanta in 1932 opened doors for Alley Pat which he never dreamed of occurring. Alley Pat was asked to make a tape to see if he could qualify for radio by a local disc jockey, Ken Knight.

The story goes . . .

"Ken Knight heard me calling bingo one night at the Lincoln Country Club and said, "boy you ever thought about getting in radio?" I said, "no". He asked me to come down and make a tape. I never went. He came back and asked me why I didn't. I was getting ready to go to medical school and I was waiting on my papers. He said come on let's go now. I went down and made a tape and afterwards I was on radio, scared as ____. That's how in happened for me in 1950 and I've been here every since. I stayed with WERD until about 1954 and then over to WAOK."

"At this time the white kids were listening to our station and there was nothing the parents could do about it. Also, the guys who got out of the penitentiary would come down to the station to see what it was like before going back home."

"We didn't have problems getting advertisers because we had

something to sell, and people were beating a path to get in on the money that could be made."

"Alley Pat has worked at 5 radio stations total, all in Atlanta. He always resisted leaving Atlanta."

"My name Alley Pat was a result of Ken Knight, the program director, being able to sell an account to Atlantic Brewery, he sold them two hours of time a day, in the morning I had Atlantic Beer and Ale and in the afternoon I had Steiner Brew, the other product. It was an hour show where I played the `low down dirty blues'. I would tell the listeners "let's get over that hang over with a nice cold bottle of Atlantic Ale", and in the afternoon I would say, "let's go into the alley and drink some beer." And the people put Alley and my name Pat together and began to call me Alley Pat. The people named me."

"Alley Pat promoted John Lee Hooker, Muddy Waters, Ruth Brown, B.B. King, and others in this caliber. The stars would come to the station as soon as they got in town to see us. They felt as if they were a part of us, not like today."

"Being a Personality Disc Jockey we had to do all of our commercials from scratch. They weren't recording and there were no buttons to push, so the spots were sold on personality because this was all we had to work with that would be original. No spots were recorded. Ken Knight and Jack Gibson both taught me how to adlib the rest of the commercials from the name of the product, the name of the product, and the address to buy it from."

"During these days they didn't plan for functions. They showed up and did what came naturally. There was a close relationship between the Disc Jockeys and Promoters as well as the entertainer, as mentioned earlier. We met them all on Auburn Avenue at the station where we worked."

"Alley Pat got a chance to work with Shelly the Playboy, Jumpin' Joe Howard, High Five from Chicago, Jock O' Henderson, and others as they came through Atlanta."

"Before NARA and NATRA the black Disc Jockey didn't have any meetings to get together with each other."

"If Alley Pat had the chance to make anything different from the past, he noted that he wouldn't change anything. We were all together and we weren't against each other, we had a brotherly love and stuck together."

Today, Alley Pat wishes that he could bring back Personality Radio.

Alley Pat seriously admits that as far as doing Personality Radio today, "I still do this, today. Personality radio puts the true Disc Jockey in the stations. I've had people in the past and present to call me and say, don't play any records now, just talk . . . we love hearing what you have to say. This is another example of the pleasure one gets from doing Personality radio."

Personality radio is exactly what Alley Pat is all about, but, he's a little shy about patting himself on the back. But Alley Pat is a pioneer for more reasons than broadcasting, in 1964 he ran for City Council, and as a result this assisted him in becoming the first black bondsman in the city of Atlanta.

As noted from Alley Pat's statement above, those low down gutsy blues were his style, but something was beginning to happen to rhythm and blues during this time.

CHAPTER SEVEN

DONNY BROOKS
He's only a 13 year old kid!

Donny Brooks

Rhythm and Blues was really beginning to take off and the earliest starting age in black radio history was being noted by a kid known as Donny Brooks.

Donny Brooks started at the age of 13. His first job was given to him by E. Rodney Jones, as he ran errands, got coffee and other go for responsibilities at WBBR in East St. Louis. Donny explains that there was one time when a radio announcer that had a show to air and wasn't able to make it, and as a result he began his career as he tells the following to you . . .

"I was asked to do his show. And because of my audience response I was given a job. My first show started out at one half hour and about three weeks later I had three hours on the air because I played what the kids wanted to hear. This was the time when Dave Dixon was the President of NATRA. I was there for about three years until he went out of business. I then went to KATZ in St. Louis, Missouri and worked there for about ten years. I was a radio announcer and worked my way up to music director and then on to operations manager. I then went to Detroit and worked with Martha Jean The Queen, and Bill Williams, Bill Crane one of the other Butterballs from Chicago, and Jack Springer that was on the all night hour. Afterwards to Philadelphia at WDAS which stood for `Why Dial Another Station?'. I worked at WDAS with Jimmy Bishop, Georgie Woods, Lloyd Fonteroy, Jock O' Henderson, Ed Bradley was the news director at that time. I left Philadelphia and went back to St. Louis. I also worked at WOBS with Larry Picus—`Jack The Bell Boy', Dave Crawford known as `The Demon', Johnny Shaw `The Devil's Son In Law.' Yvonne Daniels was also there before she left to go to East St. Louis and on to Chicago. During this time in Jacksonville this was when they had segregation and I couldn't deal with this at such an early age and I left the south almost as fast as I arrived."

"As a black person this career has awarded me with the opportunity to meet people that I never would have been able to meet as a black person. This career afforded me opportunities that I might not have ever been able to achieve as a black man."

"Highs of my career was one of being noted as music director of the year in Billboard magazine, awards from NATRA."

"Responses from the public was the most special feeling I could have ever received. I was known as `Soul Finger' and I would play the Goldfinger movie sound track and then I would do my little story and begin playing the music my listeners wanted to hear. This was when real personality was a part of radio broadcasting."

"I was a trend dresser at the time, this was the only style I portrayed as I satisfied my listeners. Especially when I worked at the only black station in St. Louis, which was what my listeners expected from me."

"At one time I used Johnny Carson's tower at the radio station in Las Vegas, Nevada which I bought that was 100,000 watts. It took me six years, and this was through the construction permit, because it was a station which had never been on the air. I wanted to do this for my three sons, but the staying power which means dollars is what kept me from maintaining this, but, I'm going to do it again. I've been very lucky and I can't complain I worked for the Star Broadcasting Company, and at this time I worked WLOK in Memphis and KYOK and WBLK at the same time, it's been a wonderful career."

"My message to the younger generation is to stay in school and get your degree. It'll all be worth it."

TOMMY LEE SMALLS

Tommy Lee Smalls took his usefulness of being a writer into higher grounds when he moved from one form of communication to another by becoming one of the legendary Disc Jockeys, known as 'The Bad Boys' also known as 'the hitmakers'.

Starting out in a store front of Savannah, Georgia; Tommy Lee Smalls became one of the legendary Disc Jockeys to penetrate the airwaves. Not only did Tommy Lee Smalls use his smooth talking talent to swoon the teens in the South, but, he took his talents North and became an immediate success as he began to control the airwaves of WWRL from 1951 to 1960. Tommy's theme of 'Quiet Village' was the soothing type of music which would relax the hard working soul during his evening broadcast, and he kept the energy flowing through the young teens blood on his afternoon broadcast. Which aired during the hours of 3:05

pm to 5:30 pm right after Alma John's Homemaker's Club. Tommy's listeners selected the latest Rock 'n Roll music and they could also dedicate songs to a friend by sending in a post card. Tommy Smalls could also be heard at night with his Dr. Jive Night Show from 10 pm to 12 midnight, during the evening hours listeners could select the music by calling in by phone and songs could be dedicated during any night of the week. Tommy's career at WWRL gave him the strong company of Alma John with the Homemakers Club, Fred Barr doing "Gospel Time", Leon Lewis broadcasting "Community News", Art Rust with the "Sports Roundup", and Doc Wheeler with the "Gospel Caravan".

Later, as Tommy's career began to reach unspeakable goals he kept in his heart the heart of his brothers and sisters in broadcasting, even when he vacationed, once while taking a vacation he called Ken Knight while he was in Jacksonville, Florida and spent time with him sharing friendship and ideas of his community concerns. Later Tommy became a very successful promoter and began to open many shows at Paradise West in Los Angeles, California which became quite a big move for he and his family, and as history tells it he became popular again in another region of the country.

Throughout Queens, Brooklyn, and New Jersey he assisted in many community service activities and later became one of the top promoters of New York. Later, by becoming friends with Mr. Shipman, (the present owner of the Apollo during this time) he was able to help those trying to make it as artists in the industry. Entertainers such as Jackie Wilson would perform at the Apollo as he was being promoted by Tommy Smalls. At one time the Ed Sullivan show in 1955 let Tommy bring his Apollo following to their stage and the local news of New York covered how many people were anxiously waiting to get a chance to see this top rated performance. Later Tommy was able to acquire 'Smalls Paradise' in Harlem which he bought from Ed Smalls in 1957. After buying Smalls Paradise he moved his show there and opened with many of the then celebrities. As a result of his opportunity to work in the Harlem Community, he later became Mayor of Harlem, and thereby became a very, very busy man. At the same time he managed Etta Jones, Brooke Benton, and the Dells, and produced many shows and records.

During the time period of 1956-57 one of the first NATRA meetings was held in the home of Tommy Smalls and great jocks such as Larry Dean, Bill Summers and Jack Walker were also in attendance.

Tommy Smalls had many friends in the industry and was able to make things happen that hadn't happened before. Sometimes these friendships would lead to business ventures and other times they wouldn't, but, the friendship was always what he considered first. One friendship involved Berry Gordy, and before Berry Gordon founded Motown, he and Tommy attempted to start a record company called VEGA, and because of Tommy Smalls having such a busy schedule the company didn't make it like Motown did, but, their friendship remained just the same for many years to come.

And the industry was never the same as the year of 1972 claimed the life of one of the black pioneers of broadcasting, who's legend still lives today through the voice of his Disc Jockey son, (Tommy, Jr.) who has the same talent and personality that his father portrayed to the community he served throughout his career.

But, before we concentrate on other individuals it is imperative to tell you exactly what was going on in the broadcasting community for people of the black and white race and how they were solved, as we examine some of the problems which were confronting them.

ZILLA MAYS
Zilla Florine Mays Hinton

Not only was the race beginning to become a part of the industry, but, it was also having it's share of problems; however, this didn't stop female disc jockey Zilla Mays from coming into an industry filled with joy and filled with problems, as she used her smooth, angelical voice to calm a listening community in times of trouble.

Better known to those in the industry and local community as Zilla Mays, a dear friend of the broadcasting industry and the community succumbed to cancer on Sept. 19, 1995. But, not without leaving a legacy for others to follow. From her days of working in the church and later with her brother's band, Roy Mays and the All Stars, Zilla was beginning to encompass a community with love.

After graduating from Booker T. Washington High School in 1949, Zilla Mays began to continue her education further by attending Reed Business College, within three years she was to begin a journey to an everlasting and enjoyable future as she ventured in to the dedicated service of the field of radio broadcasting.

As the original broadcast of WAOK signed on in 1954, Zilla began her career of mystifying the audience as she was known as, "The Mystery Lady" as audiences called in and guessed for a year who was the lady with the beautiful voice behind the mike. Zilla's charming voice swooned her audience as she continuously broadcasted to them without ever telling of her real name until she had completed a year of service with the station and only then did she decide to reveal her identity at the first concert held at Lakewood Fair Concert sponsored by WAOK, which was when everyone finally got a chance to see the person known as the "Dream Girl".

With a few years of radio broadcasting experience Zilla began to reach back to her family roots and announce her interest in broadcasting Gospel music. Thus began Zilla's pursuit of broadcasting "Cathedral of Friends" each sunday evening in the early 1950's, later to add on WAOK's first daily gospel show, which featured Zilla with sermons, and gospel music.

While Zilla quietly, and enthusiastically worked as a gospel announcer, she also gave quite a bit of dedication to those in the community as she coordinated the first WAOK radio show which reached out to the Atlanta public housing communities and gave them an opportunity to see local and national artists. Along with going to the local retirement homes and the federal penitentiary she organized Thanksgiving and Christmas food drives.

Not only did Zilla encompass the love of a local community, but, she was recognized for her achievements from time to time as she received numerous awards such as: Morris Brown College-Outstanding Achievement Award, the 1980 Pioneer of Georgia Black Radio Gold Voice Award, the 1985 Bronze Jubilee Award and the NAACP Pioneer Award in 1986.

EDDIE CASSLEBURY

Just as the 12th Annual Jazz Hall of Fame inducted Eddie Casslebury in October of 1989, others were enlighted with the knowledge of getting to know the original starter of the "Castle Rock Show", with **Eddie Casslebury**. Eddie was highlighted as follows in the middle 1950's. The press release on Eddie Casslebury read as follows:

> Although, a recent newcomer to the Miami scene, Eddie Casslebury is not a newcomer to radio. Eddie has over five years' experience, under his belt, in very competitive Negro broadcasting. Eddie received

his radio baptism at Radio Station WEDR, in his home town of Birmingham, Alabama. Eddie was a big hit in Birmingham. In fact, in a contest conducted by Radio Editor Fred Woodress, of the Birmingham Age Herald, Eddie Casslebury was acclaimed the favorite Negro radio personality in Birmingham. This was in competition with sixteen other Negro disc jockeys, who were active on the local broadcasting scene.

The understanding of his positions he held at this time and the many he would hold afterwards can be reflected with the subject of the day by an article written by David Platt, when he gathered statistics of what evidence portrayed delays, firings, and hirings of minorities in the broadcasting industry.

David highlighted his article . . . STARTLING FACTS ABOUT TV, RADIO
10-16-54
Pulse of the Public by David Platt

Alvin "Chuck" Webb. Theatrical editor of the N.Y. Amsterdam News, a leading Negro weekly, has performed a pioneering and very important service to the people of this city with his challenging series of articles revealing the appalling extent of jim crow in the radio and television industries in the N.Y. City and metropolitan area.

Several weeks ago Webb sent out a questionnaire on the employment of Negroes to every radio station and television channel in this area. He received replies from 7 TV channels and 21 radio stations. These replies are analyzed in great detail in the last article of his series which appeared on Dec. 4.

Webb's exhaustive survey showed that Negro artists are restricted to about one appearance in every 200 shows that are produced on the air which is a ratio of half of one percent.

There are 375,000 Negroes owners of TV sets in the N.Y. area. Negroes represent 10 percent of the purchasing public.

The survey showed further that less than 5 percent of the total personnel (office workers, engineers, janitors as well as performing personnel) of the radio and TV companies are Negroes.

To make matters worse, Webb points out that many local radio stations and TV channels that do employ some Negroes discourage them from joining the American Federation of Radio and Television Artists union (AFTA). Thus most of the non-union employees are paid below the AFTRA scale.

Some radio and TV outfits employ no Negroes at all. Station WMCA with an estimated total personnel of 70 employs no Negroes. Station WNEW which has almost 50 people on its payroll employs no Negroes. Station WMCM and TV channels 13 and 5 (WAAT and Dumont) refused to answer the questionnaire, thus arousing the suspicion that they employ no Negroes.

Webb turns up some more vital facts about the extent of discrimination in our local radio stations and TV channels such as:

WLIX-TV (Channel 11-the Daily News channel) employs only one Negro. Its total personnel is around 175.

Radio Station WQXR (the N.Y. Times station) employs about 90 people in all categories, but only four are Negroes and they are mostly office workers. The Times station has no regular Negro performing personnel.

Radio station WLIB (N.Y. Post station) employs about 38 people. Nine are Negroes-which gives this station an exceptional good average. All nine of its Negro employees are regular performing artists, says Webb.

NBC (radio and TV) has around 3,050 employees, but only 150 are Negroes. NBC has only two Negro artists on its payroll-Charity Baily and Natalie Hindesas on TV.

CBS (radio and TV) employs 200 Negroes—200 out of a total estimated personnel of 3,000. CBS has ten regular Negro performers on its TV payroll and one on radio.

WABC (radio and TV) employs around 1,000 but only ten are Negroes, mostly clerical and office workers.

Station WINS (Hearst) has a total personnel of around 65, Two are Negroes. Station WRL with 30 has six Negroes. Station WHOM with 70 has four Negroes.

Thus, out of a total of 8,019 people working in radio and television in the N.Y. area, only 400 are Negroes which is less than 5 percent.

Webb concludes from all of this that the color line is fading far too slowly and that the lucrative fifteen billion dollar Negro market has been taken for granted by the radio and TV companies. He notes that the part played by the advertising packaging and sponsoring agencies in the failure to make use of Negro talent is tremendous. Separate studies of their roles in the furtherance of jim crow are now underway.

Webb's survey vividly underscores the meaning of the recent statement by the Coordinating Committee for Negro Personnel that when Negroes are "omitted from the American scene" in some of the most powerful and influential media of information and education in our time, "the impression is given that Negroes have no place in American life. By the same token, if he is pictured exclusively as a clowny, a buffoon, a stereotype, we are saying in effect that he is restricted to this position in our American society."

One immediate result of Webb's outstanding series is that the Labor and Industry Committee of N.Y. branch of the NAACP is planning a city wide CONFERENCE on Jan. 15, 1955 to implement and follow through with positive program. Policy making officials from all radio and TV stations in this area as well as Negro disc jockeys and advertising and packaging representatives will be invited to attend.

As we can see, the times which Ed Castlebury and the rest of the pioneers were up against. A very negative structure was present and this structure was also reflected in many ways during 1955.

In June 1955, Eddie joined the Rounsaville Broadcasting Organization and took over disc jockey activities at WQOK, Greenville, South Carolina. Once again, Eddie was a big success. It wasn't long before Eddie was tapped to join Rounsaville Broadcasting's WMBM, Miami Beach, Florida.

The same story is being repeated again. Eddie was an immediate hit on WMBM and has developed a good name for himself in a very short time. Word has gotten around that Eddie is single, and everyday the mail brings marriage proposals from admiring listeners.

"Castle Rock" is a smooth running, lively morning show, designed to give the housewife and morning listeners a musical lift to lighten the burden of the day's activities. WMBM proudly agrees with the opinion of Birmingham and Greenville listeners and sponsors: Eddie Casslebury is terrific!

Ed Casslebury was born in Birmingham, Alabama; July, 28, 1928 and started his career in radio at WEDR in Birmingham of 1950. Ed worked as a Disc Jockey at the following stations throughout his career: WGOK—Greenville, South Carolina. WCIN—Cincinnati, Ohio. WABQ—Cleveland, Ohio. WQOK—Columbus, Ohio. WHAT—Philadelphia, Pennsylvania. WEBB—Baltimore, Maryland. WSID—

Baltimore, Maryland. WASH-FM-Washington, D.C. . Newscaster at: WEDR—Birmingham, Alabama. WCIN—Cincinnati, Ohio. WABQ— Cleveland, Ohio. WQOK—Columbus, Ohio. WEBB—Baltimore, Maryland. WSID—Baltimore, Maryland. WASH-FM-Washington,

D.C. . Entertainer Editor and Mutual Black Network News: WASH-FM-Washington, D.C. . Program Director: WQOK—Columbus, Ohio. National Black Network—New York, N.Y.:News Anchorman and Entertainment Editor. Columnist for several newspapers including TAFRIJA magazine in Atlanta, Georgia.

Eddie Casslebury's successful career has been profiled in Who's Who In Black America 1st and 7th Edition in addition to the Rhythm and Blues Month at the Smithsonian Institute in 1981. He was named Outstanding Citizen from the Alabama House of Representatives in 1989. In 1979 he received the 'Jack The Rapper-Newsman of the Year Award. In 1986 he was elected to the Alabama Jazz Hall of Fame for playing and promoting Jazz on the Air. In 1981 he was awarded the Outstanding Newsman by the National Coalition of Media Women— Long Island, New York Chapter. Elected to the Black Broadcasters Hall of Fame in 1986. Received Public Service Awards from Omaha, Nebraska; Columbus, Ohio; and Philadelphia, Pa. . Eddie Casslebury was also the first person to ever interview a very young Jesse Jackson on radio.

Eddie Casslebury was born in Birmingham, Alabama on July 28, 1928, and at 22 years of age (in October of 1950) he began his career in broadcasting at radio station WEDR in Birmingham, Alabama. Eddie worked at WEDR as a Disc Jockey—Newscaster and as his years of broadcasting matured so did his talent as an on the air Disc Jockey. Eddie's adventure in broadcasting place him with such stations as: WEDR in Birmingham, Alabama. WQOK in Greenville, South Carolina. WMBM in Miami, Florida. WCIN in Cincinnati, Ohio. WABQ in Cleveland, Ohio. WQOK in Columbus, Ohio. WHAT in Philadelphia, Pennsylvania. WEBB in Baltimore, Maryland. WSID in Baltimore, Maryland. WJZ TV in

Baltimore, Maryland as a booth announcer. WASH-FM in Washington, D.C. and Mutual Black Network News as Anchorman and Entertainment Editor. National Black Network News in New York, N.Y. as News Anchorman and Entertainment Editor. Columnist for several newspapers, and is currently writer for TAFRIJA magazine in Atlanta, Georgia.

Eddie Casslebury was profiled in Who's Who in Black America 1st and 7th edition. Other achievements and awards included: Profile of him during the Rhythm & Blues Month at the Smithsonian Institute in 1981. Named outstanding citizen from the Alabama House of Representatives in 1989. Received the Jack The Rapper award as newsman of the year in 1979. Elected to the Alabama Jazz Hall of Fame for playing and promoting Jazz on the air in 1986. Awarded as Outstanding Newsman by the National Coalition of Media Women from the Long Island, New York Chapter in 1981. Received Public Service Awards from Omaha, Nebraska, Columbus, Ohio, and Philadelphia, Pennsylvania. Elected to the Black Broadcasters Hall of Fame in 1986.

And if the above wasn't enough—Eddie Casslebury was the first person to interview a very young Jesse Jackson on radio.

Eddie Casslebury was ahead of his time and didn't know it. The talents of his broadcasting ability and his willingness to learn more and more about the industry certainly made him one of the most highest achievers in the industry.

When asked what he would do differently if he could do everything all over again, Eddie responded with, "I'd Fight Harder To Get On A Big General Market Station".

Eddie, believe us, you gave one of the best fights there was, and that's staying in there and making things happen for your community. Thank you, Eddie Casslebury for sharing your abilities with the young generation and never ending your dream, but making it better and better.

Eddie and others were fighting during this time. Fighting for the rights of blacks in broadcasting.

ED CASSLEBERRY ED CASSLEBERRY ED CASSLEBERRY
WCIN BROADCAST AND DJ'S AT MIKE 1ST MIAMI DAIRY QUEEN

ED CASSLEBERRY   ED CASSLEBERRY   ED CASSLEBERRY
MBN NEWS   MAVIS STAPLES   JACK THE RAPPER
                 MARY WELLS
                 MARTHA REEVES

[126]   Marsha Washington George

BILL WILLIAMS
HAVING FUN
WITH WILL SMITH

COWBOY BILL
CIGARETTE CONTEST

JACK COOPER

LAVADA DURST

EDDIE O'JAY
THE O'JAYS WERE
NAMED AFTER HIM

FLASH GORDON

Black Radio ... Winner Takes All [127]

FRANKIE HALFACRE   HOTROD   JOCKEY JACK
AKA
"JACK THE RAPPER"

JOE WALKER   KEN KNIGHT
ANNIVERSARY PROGRAM

KEN KNIGHT AT CBS (MAKING BROADCASTING HISTORY FOR THE SOUTH)

KEN KNIGHT AND CASANOVA JONES   KEN KNIGHT AT CBS   THE "MADHATTER"

MARY DEE

IRENE JOHNSON WARE NBPC 1ST NATIONAL PRESIDENT
PRESENTING SCHOLARSHIP

PAUL E.X. BROWN
WERD

TOMMY SMALLS
"DR. JIVE"

OPEN THE DOOR
RICHARD

BOB SUMMERISE

LUCKY CORDELL & HIS DJ SPOUSE

Black Radio ... Winner Takes All [131]

RICHARD STAMZ CHICAGO HALL OF FAME INDUCTION

RICHARD STAMZ CHICAGO RADIO SOUND TRUCK

RADIO PROGRAM SCHEDULE IN 1955
USED BY BLACK DISC JOCKEY'S
SAMPLE FROM WMBM RADIO STATION

	MONDAY THRU FRIDAY	SATURDAY	SUNDAY
6:45	SIGN ON.BLIND BOY	SIGN ON.BLIND BOY	
7:00	GOSPEL BLIND BOY	SIGN ON.GOSPEL.THON	
7:15	Joe Walker with		
7:30	Gospel request. Time. News and Weather.		
7:45	PIEZE NEWS GOSPEL BLIND BOY	PIEZE NEWS GOSPEL BLIND BOY	GOSPEL.THON
9:00	CASTLEROCK South Florida listens and loves Mr. Red, Hot, and Blue, Ed Casslebury. Ed spins the top R&B records, talks to passersby from his glass enclosed downtown studio, and sells South Florida on his sponsors. Voted top D.J. in Greenville, S.C. Set volume goes up when Ed comes on.		
12:00	MAN ON THE STREET LOCAL NEGRO NEWS CASTLEROCK CASTLEROCK GOSPEL BLIND BOY GOSPEL BLIND BOY PUBLIC SERVICE	THE PEOPLE SPEAK THE PEOPLE SPEAK THE BUTTERBALL	
1:45	THE BUTTERBALL Featuring Miami's Original Fat Daddy! 360 3/4 lbs. of Mrs. Smith, Brown-eyed Butterball on the air, in the street - nobody a Miami land mark. moves like a man, the Butterball show with Rhythm, Blue, Jive and Jazz is tops in South Florida.		SERENADE IN STRINGS
4:15	BOB UMBACH Bob Umbach spins the most requested R&B, Jive and Jazz Records, interviews. ROCK'		ROCK' N ROLL NEGROS NEWS
5:15	PIEZE NEWS	PIEZE NEWS	
5:30	SIGN OFF	SIGN OFF	SIGN OFF

Herb Kent

Herb Kent grew up on the south side of Chicago in the Ida B. Wells housing community. While growing up and acquiring an interest in radio he learned how to build his own radio equipment with spare parts from surplus World War II parts. At the early age of 16 he was accepted at WBEZ to participate in the radio workshops. Herb Kent later joined the Skyloft Players community theatrical group and quickly learned how to develop popular radio characters heard by audiences as they listened to "The Wahoo Man", "Gym Shoe Creeper", and "The Electric Crazy People".

In 1949 his first paid job was at station WGRY in Gary, Indiana where he earned $35.00 a week. Being one of the two personalities gave him the ability to polish his skills of how to produce, write, and interview as he worked twelve hour shifts each week.

In the 50's his first fan club was started and the name "Cool Gent" was coined; along with a phrase he coined, "Dusty Records"; which, described the grooves in the records which would sometimes crackle.

Herb's career has held longevity as he has worked at stations WVON, WGCI, AND WJJD. During this longevity he has had the opportunity to interview music legends; such as, Duke Ellington, Smokey Robinson, James Brown, Stevie Wonder, Diana Ross and many, many more. The city of Chicago has recognized Herb Kent more recently when they named a street in his honor, "Herb Kent Dr.". Soon after in 1995 he was inducted into the Museum of Broadcasting Hall of Fame. Not letting his audience down he moved his career higher up the broadcasting ranks as he became host of the popular dance show, "Steppin At Club Seven" while still working at WVAZ FM.

BILL "DOC LEE"

Bill Doc Lee began his radio career while at George University in New Orleans, Louisiana while working part time at a bar called the Dew Drop Inn in 1956. As Doc Lee recalls an employee who worked at the radio station there would cater to their services from time to time as he would tease him with the fact that if he ever got the opportunity he could take his job away from him and after saying this one time too many he was finally given the chance to prove his words when he was asked to come by and do one of their news broadcasts. While waiting to try his initial chance of getting a job in radio Doc Lee recalls that he was given commercials and news to read and after he completed his broadcast he left the station not knowing what the impression was until after three weeks had passed when his tape was finally listened to and he suddenly began to hear them talking about him on the radio saying that he had never been on radio before and sounded like a seasoned man after the station manager and owner listened to his broadcast. As a result, the next weekend he had a weekend job and the next week the guy who helped him get the job left the station and Doc Lee got his job and eventually was transferred to Memphis, Tennessee to WLOK which moved from Beale Street to 102nd around the corner from the Lorraine Hotel where Dr. Martin Luther King was assassinated where he worked at for one year. Finally, he moved to Chicago and worked at WBEE as head of the engineering department until an untimely car accident brought about a change in his life and put him on crutches for a while as he was once again transferred, this time to the city of Pittsburgh, Pennsylvania to assist in getting the engineering department there developed for the next six months. As Bill Doc Lee managed to get the department in order he received a call one day from Leonard Chess to come to Chicago and work at his station to observe Burke Riley doing the morning gospel show. Soon after, Burke left the station and Bill was given the opportunity to do the morning gospel show and never left the job until the end of his career in radio. As they held their ratings as the number one station and remained personality radio until the end.

Bills' attributes as a gospel morning host can be attributed to the fact that he had an open gospel show. Bills' show would one week be at a hotel, the next week at a church, the next a hotel, and various other local areas of the community. However, he does remember that at one time they scheduled an event at a hotel which they didn't know was a local gang hang out and recalls how some of the members joined in their services and were not shunned by those participating. Bill Doc Lee recalls them all as having a good time as they got a chance to hear Shirley Caesar, Albert Pugh Walker, The Dixie Hummingbirds, The Swann Super Tones, and others who have mostly passed away but made their markings in the industry.

And a most fond memory of Bill Doc Lee's is when his cousin Irene Ware called upon him while on vacation in his hometown of Mobile, Alabama to do her gospel show while she was gone and met with the President of the United States as one of his guest.

But, Bill's quest didn't stop with these types of shows as he got the chance to give a show which was sponsored by an automobile dealership, as well as, receiving sponsorship to do race results for one of the local television stations.

Loving and living through it all Bill tells of the fondest memory ever given to him in his career when he was working as a Master of Ceremonies in the crowd of 4,000 people inside the convention center at one of the local events for the Mighty Clouds of Joy when his daughter came onto the stage and hugged him after not having the chance to have seen her father since she and mother had moved away after their recent divorce. With open arms and watery eyes Bill Doc Lee grabbed his daughter, hugged and cried with her in front of everyone in the room and has never forgotten that wonderful moment in his career.

Unfortunately, as Bill Doc Lee's career was beginning to heighten even more he was downsized in his position by the new breed of radio format, "More Music, Less Talk", and admits radio has never been the same for him since this development. However, he feels that it's God's will for things to happen and there's nothing that he would do to change things in his career if he ever got the chance to go back and do it all over again.

And God's will did make things happen for Bill Doc Lee as he looks back on having the pleasure to have been in the industry with Al Benson, George Graves who used the name G.G., Larry Wynn (In Again, Out Again, Your Old Buddy Boy Larry Wynn Again); Ed Cook (The Nassau Daddy); Franklin McCarthy, Big King Cole, Joan Golden (The Golden Girl); Mohawk, Roland Porter, Magnificent Montaque who was a little over four feet tall, but when he opened his mouth, he sounded like a lion; J.J. Johnson who weighed about 800 pounds who eventually passed away, and recalls how mom Staples of The Staples Singers would cook him several pies a day and he would eat them all himself and not share them with anyone at the station; Butterball of Chicago who made a scope type antenna from the knowledge he received from his military technology which he'd learned and would put the antenna in the ground and hide his stations' location from local authorities when complaints were made. Bill Doc Lee looks back on it as a fond and happy life to share with others as he tells his story of he and his partners in radio.

CHAPTER EIGHT

JAY BUTLER

Jay Butler, a 33 year radio personality veteran, of WQBH 1400 AM in Detroit, Michigan admits that being in radio has allowed him to meet many wonderful people and he has seen many beautiful places; such as, Jamaica, Rome, Italy, London and the Virgin-Islands. Jay's career has taken him from Los Angeles, Sherman Oak, California and Detroit, Michigan. Jay Butler admits that his working for Atlantic Records, Playboy Records, United Artists and Whitfield Records has given him the opportunity to do so much.

Jay Butler's successful career can be attributed to when he was Program Director for WJLB, in Detroit, because, the station's ratings went from number 14 and shot up to number 2. Jay's ideas of format for the station and new concepts attracted new listeners and as he says, "I felt really good!"

Jay Butler's start in radio began when this happened as he explained.. "I had a friend who was working off and on with the radio station doing off and on jobs and he had the opportunity to apply for a job because one of the people who was graduating from college was leaving. My friend asked me to go down with him. He auditioned and afterwards the manager asked me to audition. The next day he came out to my school looking for me and I had left. He drove to my home and talked to my mother. He asked me if I was ready to go to work. Of course I was shocked. I finally said yes. This was in Sept. 1957."

Jay explained that in his town he really didn't have problems with critics as some others may have. Jay was in a town of about 60,000 people with about 30% of them being black, this was right before the sit-ins. Jay says, "I was doing fine and great."

Jay feels that his career has rewarded him with much more than he expected. "Yes, the places I've gone, the people I have met. I'm comfortable."

Even though he may have been comfortable there were some highs and lows. Jay explains, "My first years in radio were all up hill. From the time I got in until I went to Nashville. In Jackson, Tenn. Jolly Joe Norsley, who I thought was doing a great job. Roland Porter was leaving WJAK when I got there and was going to another radio station, as far as I was concerned he was one of the greatest D.J.'s that I had seen. Jolly Joe and I were wonderful friends and one day he left in his volkswagen after leaving my house, and he was killed in an automobile accident. I felt very sad about this and I miss my friend. I never had anything bad things to happen to me career wise, but, losing my dear friend was certainly a low which I experienced.

Another high in my career explains Jay Butler was breaking, 'Baby I'm For Real'—By the Originals, there were about six or seven disc jockeys who worked along with him in Detroit in making this happen.

A Pop record by Elton John was also broken by he and his co-workers in Detroit at WJLB.

Jay explained that his most special feeling while broadcasting over all of them was the following:

"Being able to talk people out of committing suicide, they called to talk to me and I talked to them. In Nashville this happened and lots of them in Detroit. A lot of D.J.'s experience this. Butterball and I raised money for families to help pay their rent."

AWARDS

Detroit-Disc Jockey of the Year Award from the city.
From the state I have some proclamations.
Jay also feels that it was also an award to be the holder of a picture of the staff of WJLB—which included Martha Jean, (one of the first females in radio broadcasting), and myself. Jay also attributes another personal award is a picture he has of he and Donnie Simpson, (classic because Donnie was there in the beginning, striving to make things possible for us)."

As part of some of his company promotions he wore a Superman outfit. This was his way of stimulating his fans.

Jay explained that disc jockeys did the talking during the early days and competed with the ones before and the ones after. We were trying to get that audience. We as Jocks would rap and rhyme and that's how we got the people's attention. This was very competitive. You were trying to be just as bad as the guy that was on before you. We were what you would call Air Salesmen. The name of this game was to sell time. Sell the commercials. I did one from a song name "Had A Call", and I used the record and talked about 2 or 3 mins. with Had A Call. (This was a time before coke machines existed, to give you a better understanding of what people had to work with in the cities). The 24 inch disk was what we cued with. This was a time when things were done by sometimes the same person. You did everything, traffic, etc.

Jay Butler strongly feels that the fact that he likes giving and that's what he's here for is what keeps him going. And most certainly, recognizing that God has a purpose for us in life.

When asked, "what would you bring back from the past if you could?", Jay Butler had the following to say, "The fun and the enjoyment of doing this work. The for real commorarity that's missing. The type of talent from the past doesn't exist anymore. Nobody did it for them in the past. The creativity made what we did work. They were talented people who survived of themselves. Everything started in the South. Radio stations cannot find personalities anymore. Today, people are known by their name and not by their radio personality. The evidence is there when you look at the personalities from the past.

Not only did Jay Butler become a legend for the city of Detroit, but, the team of Sid McCoy and Yvonne Daniels did as well.

SID MC COY & YVONNE DANIELS

Sid McCoy and Yvonne Daniels
One of Chicago's most talked about Disc Jockey's has been Sid McCoy and Yvonne Daniels.

Sid McCoy

Sid McCoy did a late night jazz show for WCFL for seven years beginning in the late 1950's. Sid McCoy and Yvonne Daniels teamed together in the 1960's and caught the attention of listeners from all over the country. This relationship lasted until 1964 when Yvonne left because the station changed to Top 40. This team of disc jockeys are part of Chicago's talked about history.

YVONNE DANIELS

Her name has been mentioned a lot by many in the industry. After leaving Jacksonville, Florida her career boosted when she hit the airwaves of Chicago.

She was brought to Chicago by rock station WYNR to do the overnight jazz show to compete against Sid McCoy. Yvonne and Sid developed a friendship and often called each other to chat while in the middle of their respective shows.

WYNR later changed it's format to all-news several years later and Yvonne joined Sid at WCFL. Listeners from all over the country were mesmerized by their staged 'feuds' over the 50,000 watt station and helped both Yvonne and Sid to give the station the attention needed to being one of the best.

Later in 1964 the station went Top 40 and Yvonne spent nine years doing an evening jazz show at WSDM. In 1973 she was hired to do overnights on Top 40 WLS, where she became the station's first female disc jockey. Yvonne held down the 2 to 6 a.m. shift for nine years and then left to join urban WVON in the morning-drive slot.

Later the station became WGCI AM and after being assigned to overnights once again, Yvonne left to become the morning-drive personality on soft-jazz WNUA. Yvonne was a role model for many women in radio, she was classed as, Chicago's 'First Lady of Radio' who later died of bone cancer on June 21, 1991.

Bob Summerrise

After leaving the city of Los Angeles as the first black student to attend Fremont, Bob Summerrise decided to further his career at Polytechnic Institute where he got a chance to learn more and more about music and thereby spent all of his money buying records as he practiced his ambition to become a disc jockey. Later, he decided to leave the area and visit his mother in Burlington and decided to go to the local radio station after finding himself unemployed, Bob Summerrise recalls that he was giving it his best at trying to get them to hire him. As a result his time as a disc jockey began as one hour a week, onto two hours a week, quickly, he achieved the accomplishment of three hours, three nights a week as he played rhythm & blues, and also included some jazz. With a black audience of 35,000 in the northwest during this time Bob admits that rhythm and blues was not being played, Charles Brown, Ray Charles and people of this particular caliber were not given the chance for airplay at first, and as a result the new breed of dee jays were starting to add them in. As Bob recalls the white audience was what he had as an audience because of the population gender in this area.

As Bob recalls he was able to promote local beauty salons, local stores, and later he was able to get the bigger accounts such as the beer and car dealership accounts. At stations such as KBRO in Burlington, then KRSC with Bill Apple in Seattle, as he replaced Bill Apple after he got his engineering license and completed his education.

Bob remembers his title as being called an announcer and an engineer and carried this trait with him when he arrived in the city of Tacoma, Washington where they built a special broadcast facility for him at the top of the "Burger Bowl" where he broadcasted his show from the top of the Burger Bowl as the young ones would get their hamburgers, sit and listen to him broadcast and as one of their peers would go along and collect their request, they would climb the ladder to the top and give him their request, which, Bob recalls as one of the most monumental highlights of his career.

Bob grew up mainly in Jonesboro, Arkansas and explains that his father was a postal worker and even though his parents had separated they were considered an upper class family during this time and thereby became the first family in their neighborhood to have a radio. When listening to radio at this time Bob recalls their being able to pick up reception from the areas of Chicago. In 1937, when he moved to Los Angeles he began his record collection and began to collect quite a bit of music and this when his career started.

CHAPTER NINE

EDDIE O'JAY

Not only was he an outstanding Disc Jockey, but, also an outstanding manager of one of the well known groups of all times, "The O'Jays", this was quite an interview to hold. Eddie O'Jay's dedication, determination, and honesty can be exemplified by the following interview which was written in part by Eddie himself.

"1950 in Milwaukee, Wi. WOKY and WCAN were two of the stations which many blacks were beginning to listen to. Hearing a black disc jockey only on certain days; usually weekends. Black records were called race music. Shorty Moore, a black disc jockey on weekends, became ill. Many others including myself took the opportunity to try out for the position. Big Al Benson was already the giant in Chicago at this time. The station management said they were still trying to fill the spot. They wanted someone more like Shorty Moore and interviewed other people. Later, a guy called Manny Moreland, Jr.

filled the spot. It was a heartbreaker for me not to get the job especially because I had already told all of the people in my neighborhood that I was on Radio, now, I had to go back and tell everybody that I would not be on the air. It was very disheartening. After leaving their office, on the elevator as I looked up inside, there was a listing all of the companies in the building, as I read it and noticed that there was another radio station in the building. I decided to try and see what would happen. I asked to see the program director. I refused to be discouraged, he finally came out to see who I was. We went into his small office to talk. I told him my story and he obviously saw how serious I was. From this conversation I learned that he had already spoken to—William (Bill) Killibrew, but, in his final decision, he gave both of us positions, midnight to 6 a.m Bill K. and Eddie O'Jay. We tried hard to play race records, but, at this time, there wasn't enough in the station's library. Basically, Duke Ellington, and big band music. There was a black distributor in Chicago which was where we could go to get the race music needed. The Leaner Brothers United Distributors, the first and only black distributors in Chicago. V.J. records were an intricate part of race music being distributed and they were used by stations everywhere." Vee Jay Records were the 1st all black record company on note, Vee Jay—Vivian and Jimmy Brackens were before Berry Gordy's motown label.

"During this time I didn't have a goal in mind. I had a strong desire to be a disc jockey. I had a pure interest in wanting to be an announcer. During this period there were some black disc jockeys who were trying to sound white. Most stations wanted, a black sound because they were trying to get the black dollars. I tried speaking well because I thought this was the way you should be. It was when I got to Cleveland when I came to realize it was time for me to let me be myself."

"As a result of being myself I remember having my first spiritual high in the industry. I was forced out of the habit of playing what was termed big band music, and trying to be culturally upgraded."

"My move to Cleveland came by way of a flip of a coin. From Milwaukee by train to Cleveland and staying at the famous Majestic Hotel, the only and largest black hotel in Cleveland. Staying there for a few days to get the lay of the city. I drove a cab for while trying to pursue my career in radio. Bill Hawkins was big on WSRS radio at the time. He had a remote daily show from 4pm to 7pm while sitting in a window of his record shop, and people would watch him broadcast. I observed Bill eagerly and finally spoke with him. I will never forget that Bill told me I sounded too white. That was quite a blow to me. Later I met a guy by the name of Teddy Blackmon, an old showman, doing a remote show from a busy place named "Slaughters Restaurant". A real fine looker, Juanita Hayes, Teddy, and I were. Teddy would leave everything up to Sugar Lump "Juanita", and I. We basically ran the show by ourselves, midnight to 6am. I had several jobs. I worked for a while at Crayton's Sausage firm as a salesman. Hancock candies, as an independent life insurance carrier. I also did theatrical work at the playhouse there, the Karamu House, and was fortunate to be casted in several productions. This helped me gain recognition. I worked for Dictograph Fire Safety Products, and upgraded to District Manager, and finally receiving a call that there was a station opening up and were interviewing. I got the job at WABQ. Meeting this great man, Harrison Dillard who was the Olympic Gold Winner and Selena M. Williams was going to handle the women's talk show, Harrison was the Program Director, and Reverend I.H. Gordon did the gospel show and others who handled other sectors such as jazz. This was in 1959. It was a big event."

"During this time I was beginning to develop my career more and more and the song Sweet 16 sung by B.B. King was one of the songs I broke on the air waves. I had been leaning toward the big band style and wasn't aware of it. I thought Blues music was all sadistic. People would tell me, `Eddie you were cooking this morning', which led me to believe that I was doing things right. Management was telling me something different, I wondered how could he live up in Shaker

Heights, where you have no contact with black people at all and tell me what blacks wanted to listen to, and I became very sensitive about it. The owners subtly told me that if I didn't play the format, I would be playing records at home, which meant that I would not have a job. Because of this I began to play blues immediately; such as B.B. King, Bobby Blue Bland, etc. somehow I took a liking to Bobby's and B.B.'s records. Later on, one Sunday after church, my wife and I were on our way to see a movie. Driving pass Gleason's Musical Bar there was a line around the place. After the movie we again passed Gleason's and noticed the sign saying B.B. King. We parked and the owner's son beckoned for me to come in. Before we had gotten to the center of the club the music stopped and B.B. announced that there was a Disc Jockey in the house responsible for making his record known in Cleveland, and this was my first high in radio."

"My next high was when I won a convertible Sting Ray from Italian Swiss Colony Wine. Nat King Cole did the commercial for a wine called "Ariba". 4 radio stations was to compete in selling the largest quantity of Ariba on The Air, they were in Detroit, Pittsburg, and Cincinnati, and Cleveland. The Top Air Salesman would be the winner. I made "Ariba" cocktails every where I went. After shows at Karamu House, at cast parties I would mix ginger ale and Ariba in a big bowl. People called me, `O. Jay Ariba' when they would see me driving down the street in Cleveland neighborhoods."

Making of the O'Jays

A group known as the Mascots, from Canton and Massibon, Ohio came to perform at my hall, called The Call & Post Ball Room. These five fellows were well dressed and immaculate. Hair styled and all, they requested a spot on the show. And we had a spot open, because one of the artists didn't show. I put them on as the show opener. The audience exploded. I played them many times. They continued to burn the house down. As time went on we had a discussion about Management affairs. In our discussions we spoke of name titles. If I'm not mistaken, it was Walter who said I think we should be called the O'Jays and the rest agreed. Being proud, I asked, why? They replied because of your integrity and popularity. We had a five year contract, we all signed . . . '(The O'Jays), Eddie Levert, Walter Williams, William Powell, Bobby Massey, and Bill Isles. We tried to get bigger promotions and recording contracts, my pal and good friend Berry Gordy at Motown made an offer, but we were able to get a better deal with H.B. Barnum, which was to take the boys out to California. Later, I blew my job at WABQ and H.B. Barnum took them on and by the time the O'Jays got back East the contract was expiring. Also the FCC was deciding whether or not Disc Jockeys could play records of the groups that they were part of which of course put my Management position on hold with them. I couldn't do anything with them at this time and by the time the FCC reached a decision stating that they were concerned with Programmers more so than Disc Jockeys, my position as their manager was made more clear and that I could renew the contract again, but it was too late, they had already gotten another contract signed with another person. Of course there was a toss of the name, but we all decided to let them keep the name and we all remained close friends.

"Years ago. As a rule of thumb some stations kept the names of the disc jockey which aired for them, such as, "Dizzie Lizzie" and "Okee Do' Kee", their names belonged to the station. They could put the janitor in and he would be Okee Do' Kee. Also you couldn't take a job within a fifty mile radius if you left a station. Black jockeys had to move every time because of this contractual agreement and it may still apply at many stations today."

"Because radio was the important part of the community at this time, we had to be creative with our personality radio style. The audience would get to know you and to love you. I would go to pool halls, dance halls, everywhere in the community. And as a result of this became popular. What I always say to our young ones who are getting involved in radio today, is . . . be active with the community, music along is not going to do it alone. God, and the commitment of what you give to the community is what will keep you going."

"In closing, to give a message to the younger generation to come. Remember your heritage. I know for a fact we're not playing enough of the music from our heritage. We have artists who are starving to death, especially blues artists because we don't recognize them as much as we should today. Europeans accept our musical heritage and it is very disheartening to see us give it up. Sam and Dave's music may not be what we enjoy, but, somehow a person who covers the songs of Sam and Dave's gets widespread recognition. Because we don't accept our own heritage. It's up to us to keep our musical heritage a part of history."

CHAPTER TEN

LUCKY CORDELL

Not only do we have someone by the name of Mr. Lucky in the many files of broadcasting history, there's also someone from the windy city of Chicago by the name of Lucky Cordell. Lucky's career began when he came out of the army and talked with a friend of his named Bill Gayles who worked down at Ernie Leaners Record Distributors. Bill was going to school at Radio Institute of Chicago who had said to me to come down to the school and sit in on a class. As a result Lucky became a student at the school. Lucky Cordell's broadcasting career began by attending the Radio Institute of Chicago as he completed the four year program in three and a half years. Afterwards he taught for six months at the school, but, he wanted to be an on the air disc jockey and left the school to began his career as a disc jockey in 1950.

Lucky took the techniques which he learned from the institute, radio drama presentations and taped speech practices which were recorded when you entered the school and recorded again when you were ready to graduate and as a student he and others would compare the difference, and through this training he discovered the talent and valuable training he needed in order to make it in the broadcasting industry during the 1950's.

As Lucky Cordell began to apply at the different stations in Chicago, he found out quickly that blacks were not being hired by the white stations but he tried them anyway. Even though Lucky received discouragement from Sam Evans it didn't keep him from wanting to reach his goal of becoming a radio deejay at the station. Finally Lucky went to see Al Benson at WGES which later became WYNR and after several meetings with him, he was hired as a substitute disc jockey to replace him when he didn't feel like working, later he was given an hour every night and remained in the position for a couple of years.

Before Lucky's arrival to Al Benson's station, Al Benson had been known as the top dog in the area. Representatives of big companies such as Pepsi Cola and Coca Cola were always knocking at the stations door because this was their avenue of reaching the black community. This was a very powerful status for a black man during this time in history and Al Benson loved it all. Sales were rocketing sky high and if you wanted to get your product bought Benson could do it for you, but, as time went on Al Benson hit a slump in sales and decided to recruit others to assist him. Lucky Cordell and another disc jockey by the name of Tom Duncan were hired at first to work with his newspaper and to do the sales to show management and the owners that these young, hot shots were trying to do the same thing I'm doing and sales just weren't where they should be. Later, Lucky was able to sell his first account and gave the contract to Al Benson to sign and as a result received permission to sell and sign contracts. From this Lucky received commission and was able to sell many, many more accounts for the station and Al Benson. However, there were some clients who did not want Al Benson to do any of their commercial spots because they wanted a younger, fresher announcer to do their spot or they would not make the sale. Al Benson didn't recognize the problem and decided to do one customer's commercial spot and the client called the station and spoke with management and complained that they wanted the person who they sold the contract to, to do the sale. They refused to pay for the commercial and the owner called Lucky in and reviewed his past commercial spots with him. Unfortunately, this caused a problem for the relationship between Lucky Cordell and Al Benson, and resulted in Al Benson releasing Lucky from his services as an assistant. Because of this, Lucky decided the next morning that this was his opportunity where he could finally go up stairs and talk to management about furthering his career, because in the past he wsn't allowed to go up stairs because he was under Benson's payroll.

Lucky remained with the station a little longer but left because all of the credit was given to Al Benson and all of the air time. Other disc jockeys such as Franklin 'Sugar Daddy' McCarthy, and Tom Duncan worked under the auspices of Al Benson and later others such as Richard Stamzs, Jessie Owens 'the Olympic Runner', Rick Ricardo, Russ Vannoy, who were also Disc Jockeys at the station and at this time management was beginning to hire more disc jockeys under their contract, and not Al Bensons'.

Because of the management style at the station Lucky Cordell decided to leave and worked at WGRY now known as WLTH in Gary, Indiana which became a very successful career opportunity for him as he broadcasted gospel and rhythm and blues. Lucky's broadcast was dedicated to an approach which he did unintentionally and was receiving applause for. The audience listened to him as he did the gospel show with a totally different movement—softer tone, and a smoother approach, he would then play a few commercials and then began his rhythm and blues show. Because of this people thought that he was two different people. Once Lucky was told by a lady after church that he wasn't Lucky Cordell and that he should be ashamed of himself being an imposter because she listened to him on the radio. He thought this as being hilarious and still smiles at this and other effects he and his fellow disc jockeys had on their audience. After eight years in Indiana, Lucky asked for a raise and was refused. The next day Lucky went back to Chicago to WGES where he started and asked if they finally had a slot for him. WGES hired him back and he went back to his own program.

Later, the station was sold and became WYNR, also known as, "WINNER", and this became the first time in Chicago that they used format fast radio at a black radio station which was in 1958. The new owners kept Lucky Cordell and Roy Wood and let everyone else go. The new owners were looking for a sound which would be acceptable to both markets, black and white. This worked successfully, but, later the owners decided that the power which they had in the black community could be changed from 30 % white and 70 % black and become 40 % white and 60 % black in order to get more power over the white community, which they finally did and later they went about 50—50. Which began as a lost for them with the black community. The black base was getting lower and lower for them. As a result a little black 5,000 watt station on the river known as WVON was started by Italian Leonard Chess and played all black music, and had all black announcers and was on the air 24 hours a day. At WYNR, used Dick Kemp, a white announcer, known as `The Wild Child' and others to get the white audience and they were getting all the attention they needed, but WVON snuck in and took the black audience from them. This took the black audience from them and very little white audience was left, as a result they turned the station into an all news station, and maintained Lucky Cordell and Roy Wood. The all news station didn't work very well and the owners finally sold the station. Unhappy with the situation, Lucky went over to WVON since Leonard Chess had already asked him to come and work for him, and even though he had been trained for news, he enjoyed being a disc jockey more. Lucky started as a disc jockey at WVON and later became Assistant Program Director, next Program Director, Assistant General Manager and finally General Manager. Upon leaving WVON Lucky was promoted to Assistant to the President of the entire corporation. Lucky saw this final position at the station as an honor but because he would not change from doing the public service work and play only music he was asked to come downtown and asked to fire a couple of disc jockeys; asking, "Who did they have in mind?" they replied, "they didn't care", and they also felt that they had too many high paid disc jockeys. The owners had six other radio stations and were paying them less at the other stations and worked them more hours. Lucky felt that a situation such as three hours would bring a disc jockey on smoking and adding too many more hours would cause either his beginning of the broadcast or end of the broadcast to be less than the perfection they always tried to give. Because of Lucky Cordell's refusal to unstaff some of te disc jockeys, three months later he was given the Assistant to the President position and without consulting him they made a big news announcement about the promotion and moved him downtown to 1 IBM plaza and as time went on they put somebody in who could do their deed.

Later on during the day after the announcement had been made all of the disc jockeys but one came to Lucky's house and met with him to let him know that they were on his side, and if they needed to strike they would. Lucky explained that they can't strike because 'Lucky Cordell' was promoted, but the best thing to do is to go back and stick together, and when they come to fire the least of you i.e. the part time, and weekend man, this is where you would need to stick together. As time went on these disc jockeys were released one by one and because of mostly, financial situations they held onto what they had until their career move was decided for them or they decided to make a change. Lucky Cordell resigned and after a long discussed meeting with the owner he decided to end his career with them. Lucky asked for one years' full pay from his two year contract which was left. Lucky agreed that he wouldn't go to the NAACP, Urban League, etc., or go to the media on them, (even though he had no intention to do this), they felt he would, and so Lucky gave them his word and shook hands with them. Later that afternoon it was announced that he was resigning and Lucky responded to everyone that yes he was happy with this decision.

Now that enough time has passed Lucky feels comfortable telling about the above experience.

Soon after, Lucky went into the record industry and when the economy dropped, the record industry couldn't afford new artists and gave him many creative challenges to make things work in this industry. Lucky admits that this position was quite an enjoyable one for him.

Not only was Lucky Cordell active in the local community, but he was at one time President of the National Association of Television and Radio Artists. When he worked with NATRA he found that some of the expenses could be cut and made a few changes so that the organization could survive. At one time NATRA was $80,000 in debt and as a result of changes made under his administration the organization was out of debt and was able to survive and maintain it's status. This association was made as a networking tool for blacks in broadcasting especially the disc jockey so that they could assist each other in the industry and also share ideas for themselves and the black community. Many times Lucky Cordell met with one of the original thirteen, 'Ken Knight' at the conventions and as a result remembers him as a kind gentleman. Even though Lucky Cordell is no longer on the airwaves, he still believes that radio is only effective when the listeners are satisfied with their local broadcasts, and strongly feels that record programmers cannot be successful because of the demographic differences, and when you eliminate personality the only thing you can do is play records. Lucky Cordell was a strong factor in the field of broadcasting and because of his dedication and service shall be remembered as such.

CHAPTER ELEVEN

JOE WALKER "THE BLIND SCOTSMAN"
Joe Walker

Telling It Like It Is . . . and as he see's it.
Being blind hasn't stopped him at all!

Joe Walker Sports Commentary. Daily. Monday through Friday. For all you sports buffs, Joe Walker tells it like it is.

Since Joe Walker started in radio in 1954, he has been a Sports Director. Beginning at WMBM in Miami, he stayed there for 13 years, before joining WAME in 1967. Joe moved to Atlanta in 1969 and immediately landed a job as Sports Director at WERD (currently WAEC), staying 9 years. In 1978, at WRNG, Talk Show Host was also added to his title of Sports Director. He spent 2 years at WRNG before accepting a position at WAOK in 1980.

Joe Walker's awards are many, including the Pioneer Award from the Atlanta Association of Black Journalists, and Sportscaster of the Year for 1977 from the 100 % Wrong Club.

For many years, Joe has been a member of NAACP (since 1954), and SCLC (since 1975). And, not only is he a member of the Church of Incarnation, but Joe Walker works with his church's little league team. As part of his ongoing effort to help his community, his participation in the Feed the Hungry Telethon on WGNX (Channel 46) did not go unnoticed.

Joe Walker loves his work, but especially likes meeting sports personalities. In his spare time, Joe enjoys bowling and reading. In whatever he does, you can count on Joe Walker to be the best!

Blind Sportscaster 'sees' action for radio listeners
By: Doug Carlson—staff writer
7-28-86

As a blind 7-year old second baseman for the Florida School for the Deaf and Blind's baseball team, Joe Walker liked to imagine he was playing for no less than the New York Yankees. Although he didn't consider a career in baseball ("I couldn't hit, but I could fight and cuss"), he did dream of a career in athletics—as a sportscaster.

One recent Monday morning at 10:45, on the 10th floor of the Renaissance Square building downtown, Walker, 53, is in constant motion. Head rocking, hands moving, the Atlanta sportscaster talks about the Evander Holyfield-Dwight Qawi championship fight as he dubs tapes for a 12:20 sports report on radio station WVEE. Never mind his blindness; Walker "saw" the fight from the press row.

"I had Qawi slightly ahead after 12 rounds, but the last three rounds made the difference," he says, spinning to his left to rewind the tape. "Holyfield quit standing there and started jabbing like he should have been doing all along." Walker could hear and feel punches being thrown and knew from the crowd reaction how hometown-favorite Holyfield was doing.

Since becoming a sportscaster in 1953, Walker has "seen" plenty of fights. "The Holyfield-Qawi fight was a good one, but I wouldn't say it was one of the 10 best fights of all time," he says. "The second Ali-Frazier fight was the greatest fight I've ever seen."

Walker speaking into the microphone and in a deep, fluid tone, says, "1,2,3,4 . . . this is Joe Walker . . . Evander Holyfield said Dwight Qawi fought like a champion in Saturday's, ah that's no good."

Four takes later, he's satisfied. He dubs in quotes from Holyfield, pops the tape out of the machine and delivers it next door.

"I've been here six years, five months and 12 days," Walker says, "I've done everything I've wanted to do. I've been to NBA games. I've seen major league baseball. I've been to championship fights. I'm comfortable with my life. In fact, if I had a choice, I'd rather stay blind than be given vision."

Walker doesn't come across as being handicapped. He is blind the way some people are short or tall. Feisty and confident, he speaks like the tough person he's learned to be.

Born in Miami, Walker went to the Florida School for the Deaf and Blind in St. Augustine when he was 6. "When I first went there, the blind kids were tough. They sensed how nervous and scared I was and said, `Come here, and I'll kick your butt.'"

A boy named Ray Charles made life particularly hard for Walker. "Ray Charles kicked my tail—breakfast, lunch and dinner for three months," he remembers about the now-famous musician.

Walker eventually learned to fight back. "You think Ali and Frazier fought? One day, me and Ray fought so hard the whole class had to leave the room."

One year and many fights after his arrival, school officials thought making Walker, a second-grader, and Charles, a third-grader, roommates would force them to become buddies. It eventually worked.

Ten years after the two scrappy blind kids first roomed together, Charles' mother died and he quit school to become a professional musician. Walker remained in school until he was 21, earning a high school diploma. He considered going to Florida A&M University. But he chose to find a job instead when the university's president told him a blind man would never become a broadcaster.

Walker found jobs in radio scarce. "They'd all tell me, `We don't want any blind, black kid around here.'"

But when a friend became ill and couldn't do her radio show in Miami, Walker filled in and subsequently earned a job at the station, where he played gospel music and covered the Brooklyn Dodgers during spring training. Since then, he has worked at stations in Miami, St. Louis, New Orleans, Houston, Louisville, Ky., and Dallas before coming to Atlanta in 1969.

Walker first dreamed of becoming a broadcaster when he was a boy listening to Friday night fights on the radio. "I'd hear those voices in the radio, and I just wanted to crawl inside of it," he says.

Now his voice comes out of the radio two times a day on WVEE and four times a day on WAOK. He writes his own Braille script for three daily live shows with help from Carolyn, his wife of eight years—"eight years, five months and 23 days"—who reads wire stories to him.

On Sunday afternoons, he does a live talk show on WGST. "Hell-raising time," he calls it, in which he interviews everyone from politicians to sports celebrities. He regularly covers the Hawks, Braves, Falcons, college athletics and anything else of significance in sports. At Hawks games, the squeaking of shoes helps him follow the action, and he sits behind play-by-play man John Sterling near the bench to hear what's going on.

Walker, who has three grown children from a previous marriage, will take his 5-year-old son Joseph—already an avid Dodgers fan—to see his first baseball game in August when the Braves play the Dodgers. If afterward Joseph says he wants to become a major-leaguer, he'll get no discouragement from his dad. The man knows stranger things have happened.

HERB LANCE

Known as the man who always answers you, "Oui!" Herb Lance became a disc jockey in 1960 after he had already become known as a soulful singer as he sang `Close Your Eyes' and `My Kind of Girl' in 1938 and 1939, while doing this he got the opportunity to work with Dizzie Gillespie and Quincy Jones. Later, he wrote the song, `Mama You Treat Your Daughter Mean' and still receives royalty payments from this song. While he stills remembers the likes of Ken Knight who always reminded him of someone who used his thoughts to get things done the right way. Remembering the fact that Ken Knight took WERD station number to Jacksonville with him in order to keep the encouragement of black ownership strong. He remembers that Personality radio was still around doing this time and up until the time of automation becoming a part of the main stream they were able to keep the personality in radio programming.

During this time money was tight for everyone and the cost of getting married had to be planned, and planned very well. Herb Lance strongly remembers being married by Rev. Martin Luther King, and recalls the fact that he only had forty dollars and made the mistake of giving half of it to the Rev. King, and saw that he had made a big mistake by not having broken the twenty into change to pay for the wedding. Later, as he met Rev. King at Paschals Restaurant while going there for dinner, he recalls being asked how he and his wife were doing. He smiled, and said everything's ok, but, he was still thinking about the large amount of money he paid in order to get married.

But this didn't stop him from having more fortune in his life and happier memories, because the one that made him smile the most was when someone called in and said that they were Max Cooperstein calling from Chicago and as he was on the air he was told that they were having a big convention in Chicago, as the caller named all the stars who would be there. As he quickly proceeded to interview the stars such as Ella Fitzgerald, Lou Rawls, and others of the same caliber, he spent 30 minutes interviewing the stars, and as he spent this time doing his broadcast diligently to his listeners someone called in and told him he was talking to George Kirby who was playing a trick on him. He quickly said, "George Kirby", and they both began to laugh. George Kirby was one of the great impersonators of all times and he had pulled one over on Herb. Other happier memories was meeting Muhammad Ali, who was surprised at the way the broadcast was done from a black station, and he told him they could do this type of format because it was a black station, and they talked for awhile and then Ali spoke to the listeners. Another fond memory was not only later getting to broadcast in the same city with Zilla Mays, but, also sang with her at many performances before this time as they toured all over America.

Working at WERD, and WIGO, has never given him a disappointing time in his career, and today he still uses the personality he possessed as disc jockey as he operates Revelations Gospel Music store.

CHAPTER TWELVE

BILL WILLIAMS

In high school Bill Williams developed an interest in broadcasting.

At the private school which he attended he and a few other students that shared the same interest were allowed to broadcast for their school. This school in Martin County, Fla. allowed them to put horn speakers on top of the school building, and broadcast from the little studio inside. This was Bill's first introduction to radio broadcasting. From high school Bill went to college for a year and then he enlisted into the service.

As Bill began to learn other trades, things were developing for blacks in the industry. For one, black air personalities were beginning to become a reality. During the late 40's or 50's whites were beginning to realize that money could be made by using the influence of having blacks as station disk jockeys. After leaving the military Bill became a truck driver and one particular night he and one of his buddies was listening to a local disc jockey was doing his best to broadcast a commercial. Bill boasted to his friend that he wanted to go to the station and see if he could get a job doing what this disc jockey was trying to do. The station was offering a job for 1/2 day on Saturdays and he asked the lady who was working there if he could get the job. Bill was asked if he had any experience and he told her no. The lady proceeded to give him an application to fill out and he did just that. Afterwards, the Engineer took him into the studio and gave him a couple of commercials to read on tape, which he completed and was told to call back at 3:00 Wednesday because the manager wasn't there.

Bill Williams never thought that he would be able to call on Wednesday at that time because he would be working at the company he drove trucks for; but, July of 1957 would prove to be a winning month and year for him. Fate had it that Bill was in the shipping clerks office at the time he was to call the station and when he looked up at the clock he remembered that he had to call the station. Bill asked to use the phone and when he talked with them he was asked to come back out on Saturday. The manager of the station at the time was Joe Star. Joe asked Bill if he thought he would like radio? Bill's reply was, "Sure, why not?" Bill was given all of the station rules; such, as getting along with people and so on. Bill understood and accepted his new job, and started to work the next week.

At the time minimum wages were approximately $1.00 a hour and this was he was offered as his starting salary. At this time Bill had a lot of learning to do. Bill's experience with his high school couldn't match everything he needed to learn here; but, he went in with a very open mind. The station would open around 5:00 am and he had to make sure that he knew everything, and made it his duty to get there a hour before or stay a hour after in order to learn everything he needed to know. Learning programming logs wasn't a problem since he had done this particular job in high school; but, the challenge of learning switches and other equipment his mock station in high school had wouldn't be enough to challenge his new found career. Bill took advantage of learning everything he had to learn and appreciated the thoroughness of the training which he received from Bob Roundtree; also called, "The Bob Cat".

Bill worked part time for six months on weekends. When he would arrive at the station the person leaving from their shift would explain things to him and leave him there to manage the show. The hours would begin to vary for Bill and he would learn just what scheduling was all about. From Dec. through January he carried the sundown music from 2:00 to 3:00 pm. Bill was very satisfied with this and never knew that these slots meant something very important to others who had been working there longer than him. The midday spot was 9:00 to 10:00 am until the gospel spot, and break, one hour afterwards there would be the work shift show. For 2 to 3 weeks he carried the work shift show. One evening he was asked to work the am show. At this time he had no idea that the am slot was the most important slot on the station. Bill accepted the slot, and of course he soon found out that this was the most prestigious slot for any personality to have. During this time the share of listeners was starting to rise for the station, and it got as high as 67% share of the radio listeners from their broadcast area. This was the first time that this had been done and it's never been done again. Bill Williams was out to do the job and he gave it all he had.

One particular weekend Bill met Lewis from Detroit through one of his friend's aunt, and he was told that there was an opening in Detroit. Lewis told him to send a tape to Joe Howard in Detroit at WJLB. Lewis didn't know that Joe was working at WCHB. From the time of Lewis telling Bill to send the tape to the time of the tape's arrival Joe Howard happened to be leaving WJLB and was starting to work for WCHB. When Joe got the tape he sent it over to George White and after they listened to it they recommended him for the job. George White let Dr. Bell listen to the tape and after listening to the tape Dr. Bell arranged for Bill to be flown to Detroit over the weekend for an interview.

This was good timing for Bill because he had been laid off without pay at his home station and he gladly accepted the job offer. The pay in Detroit was $100.00 a week, and he was only making $65.00 a week in Tampa. Bill gave his present employer a two week notice; although, they said it wasn't necessary, and began to work for WCHB in 1960 and remained there until March 1973.

Coming to Detroit was exciting for Bill and it was quite an eye opener for him, since, he had never seen so many blacks controlling a station. This was Bill's first time seeing a black Program Director, Black Engineers, and Black Managers. At this point Bill realized that a whole lot could be done by blacks in the communication field. There were one or two whites working at the station; but, all of the D.J.'s were black!

Bill worked the 7:00 to 10:00 am show and later the 2:00 to 4:00 show to go up against Joe Howard who had left their station. Joe Howard didn't air until 3:00 and this gave Bill one hour on him. Ratings were very important to the station and Bill's broadcast and Joe's was compared to a lot.

There was quite a bit of diversity of coming from the South to Detroit. When Disc Jockey's broadcasted in the South their programming would be from a range of Rock to Gospel and then back to Rock Music. In the North Gospel music may have started somewhere between the hours of 5:00 am to 7:00 am and from 7:00

am to 10:00 am Rhythm and Blues would be played; afterwards, it would be the 'Tennie Weeny' time from 10:00 am to 10:15 am broadcasted by Trudy Haynes out of Philadelphia, Pa. From 10:15 to early afternoon the Big Bands would be played, this would be your Duke Ellington, and Frank Sinatra style of music. From 1:00 to 2:00 nothing but Jazz would be played and from 2:00 to 4:00 Larry Dixon would hit the air waves with Rock and Roll, and until sun down the music of Ray Charles, and others of his style would be played. This type of programming would go on until radio took a turn to the 24 hour format. This was the programming style which D.J.'s were starting to pattern after and are presently beginning to revert back to. It's nothing new, it's something listeners are crying for and the stations are willing to give them what they want . . . Quality programming for listeners everywhere.

In 1963 Bill Williams was offered the position of Program Director and he declined both times, until one day he finally accepted the challenge. After accepting this position radio took a drastic change and everything was taken out of the format of what everyone was used to except for Gospel and R.&B. . Gospel would go on between the hours of 4:00 to 6:00 am and the basic R.&B. would begin to air. Surprisingly, the public accepted what was being done doing this time. Black Disc Jockeys were looked at with a sense of pride from the community, and there wasn't much said if any, and for those who might have thought about saying something, they really didn't know that they had the power to do or say anything about what was taking place.

Air play was the decision of whatever the Black Radio Stations wanted to broadcast. As time went on Pop Music began to take over the air waves, and many listeners were requesting this type of music; but, it was still the decision of the station. This did bring about a change; however, because the play lists began to get tighter and tighter, and finally personality radio phased out everything the pioneers of Black Radio were used to doing.

In 1967 during the time of the Watts Riot taking place, Bill was beginning to look for something different in his career. Bill left the position of Program Director and began to work for Stax Records in 1973, this lasted for a span of two years when Stax Records went out of business. From 1975 to 1977 Bill was unemployed and worked various jobs until getting hired with ABC and on to MCA in 1978 and retired from the position of A&R in spring 1993 to open his own company Champion Enterprises to further enhance his abilities of Management, Marketing and Promotions.

The salaries for these positions were not the highest salaries some people would want to talk about. In order for someone to be in this business they had to really love what they were doing. For instance, Bill's salary only peaked $15,000 which was doing 1972-73, and he later found out it was the sales department that made the biggest salaries.

The position of Disc Jockey, and Program Director is what Bill Williams is all about when it comes to the history of our first pioneers of Black Radio and the numerous Community Service Awards, Public Service Announcements, Medal for the city of Detroit, The Conrad Mallock, and Jack 'The Rapper' Gibson's Black Radio Hall of Fame are just a few of the honors bestowed upon him as he proudly, and earnestly worked for the public and gave them what they wanted to hear when they turned the dial each day to hear his wonderful voice welcoming them to another day of listening pleasure.

LEON LEWIS

The article reads: 'Harlem Finally Gets Justice', but, maybe it should have read: 'Harlem Gets Justice, Again'. Remembering the previous efforts of others in the industry this is another milestone for their community.

Harlem Finally Gets Justice

Thanks to the hard work of Disc Jockey Leon Lewis, Harlem achieved something from the industry which it had never received before, and due to his efforts in 1966 Leon Lewis achieved for his station what it earned, 'The Peabody Award Citation" for his station, WLIB. This was quite a milestone for Leon Lewis, who was now the Ombudsman for WMCA's 'Call for Action'. At WLIB Lewis would host an early evening talk show of broadcasting (Community Opinion) in which residents of the ghetto could call in and voice their opinions and grievances which gave them the opportunity to share with the under publicized community services available to deal with specific issues and concerns of theirs.

Lewis's involvement with this endeavor was one of true spirit as he became very involved with members of the legal, educational, and political arenas to assist him with the needs of the show. The members of the community were then able to voice their statements to those in command, sometimes, leaving them in nervous stupors trying to answer and solve their questioned issues or concerns. As this would become a more and more available service to citizens of the community, Lewis's guest speakers would get no assistance from him as he quietly sit back and let them run the show, only, on occasions would he interrupt and close out with a philosophical statement bringing the aired confrontations to a close that could fairly satisfy everyone involved.

Lewis's commitment to WLIB was surely to be admired, especially when one would read the words inscribed on the Peabody citation which read, ". . . gave Harlem a safety valve." By WLIB's permission of airing such a show and having a powerful representative such as Leon Lewis to promote it radio programming became a great power for establishing Community Opinion which permitted the citizens of Harlem to voice their feelings, through a hot line telephone interview which covered not only the negro community but the entire city.

In one interview in 1962 Leon had this to say about his callers, "The people who call up are angry, confused, or in trouble... the theory behind the show is that many of the frustrations in our community arise because people simply don't have the information necessary to solve their problems. We try to feed them information, especially advice on how to use the resources of the city, state, and federal government for their own benefits."

On any given day of his broadcast Leon could be heard discussing any subject from alimony payment problems to an argument with a Black Power Advocate expressing their issues and concerns and the reasons why he thought Negroes should fight for their country in Vietnam. Of course he realized that there was nothing he and others at the station could do to directly to assist their callers, but, sometimes just their talking with each other made his listeners feel better.

For this direct attempt to communicate with the community WLIB's involvement was recognized and was awarded the 1966 Peabody Award for outstanding local radio education which was enough proof for it's charter.

Leon Lewis's commitment to the community was already there since he was born not far away in Troy, New York in 1917 and had worked in various radio and journalism positions. His career has included Salesman for an Albany station and circulation manager for the Amsterdam News. In his early years of working in the broadcasting industry, he wasn't able to get a successful career in radio at first, but he found a ready career on the ethnic air waves.

After achieving success at WLIB, Lewis joined WMCA as Assistant Public Affairs Director in 1967. Throughout the years he received prestigious awards in journalism which later led him to a teaching career at Fordham University.

We all must applaud Leon Lewis for his mission of achievement in the industry and remember what his main goal was as he quoted: "The Walls of the ghetto are so high, that the people can't see in and they can't see out. I'd like to see those walls come tumbling down."

As others tried to get more information out to the community, so did Leonard Evans in 1954.

We still salute you Leon Lewis and thank you for the opportunity to let the voices of the community be heard.

LOUISE WILLIAMS BISHOP

Although, Louise Williams Bishop has been in radio for more than 36 years, I must say that she is quite a woman. Louise sent a copy of the following concerning her career, past and present.

Louise Williams Bishop
State Representative

For the past 36 years, Louise Williams Bishop has been hosting a radio program 6 days a week. In addition to that she is an ordained Baptist Minister and is able to enjoy working with "This gift from God", nights and weekends. Louise Williams Bishop was elected to the House of Representatives in the Commonwealth of Pennsylvania and at interview time was running for her 3rd term.

JOHNNY ALLEN

Johnny Allen of New York, for many years had the only black school of broadcasting. Johnny Allen concentrated on people of color who hadn't had the opportunity to become disc jockeys and since 1969 he has steadily maintained his goal throughout the United States, and Trinidad. Johnny Allen was named "the Duke", by those who came through his school because of his mannerism of instruction. Some of Johnny's highs have been working with the community. The low of his career was when he worked with someone who tried to keep his self esteem lowered which was a turning point in his life when he reached within him self and decided to rule out the negative factors from keeping him down. Johnny Allen has also received from a city award from the city of New York at the same time Eartha Kitt did. Johnny Allen's goal has always been that of trying to take the talents of the pioneers and use their personality traits to get the talent portrayed to the younger disc jockeys of today and tomorrow which he trained magnificently.

CHAPTER THIRTEEN

AL JEFFERSON
BUTTERBALL

Al Jefferson

Al Jefferson can be remembered as one of the radio and record industry veterans. His long and successful career has included working at stations, WMID & WLDB in Atlantic City, New Jersey; WOOK and WUST in Washington, D.C., and WWIN in Baltimore, Maryland. Al Jefferson later left the broadcasting side of the industry and began another career as an independent record promoter whereas he provided services to Atlantic, Capitol, and Island Records.

Butterball
(At the Wammy in Miami)
WAME in Miami

Butterball

Butterball as highlighted earlier on the miami program chart did significant things such as going to the detention center in Miami and took the kids to the beach for five years each Sunday, he also bought them food, and cooked it himself. A lot of kids that are now politicians and school teachers have him to thank. Butterball got some of the black neighborhood streets paved and street lights in portions of Miami. He did real good community service work and was strong in promoting the community with black radio. He played Aretha Franklin first down in Miami and others picked it up from this point, this is where her music air started.

There were other Disc Jockey's named 'Butterball', but this one is from WAME in Miami, better known as (WAMMY in Miami).

DOUG STEELE
Coast to Coast . . .
It's Doug Steele

Broadcaster, program director, and syndicated radio show host.

Host for the nationally acclaimed syndicated radio show, Soul Coast to Coast, based in Atlanta, Georgia. Doug Steele has enriched the knowledge of many Americans with facts on recording artists. Soul Coast to Coast is a quality show.

When Doug Steele began working in radio during the latter years of the 1950's there was beginning to be a trend with the establishment of many record companies. Especially in Chicago and the Mid-West regions. A lot of major record companies, Columbia, RCA, and Capitol wouldn't have a black artist on their label. It was the other side labels that would carry all of the black artists. At this time artists were trying hard to sell and they would, even if it meant sounding like their white counterparts. One organization by the name of ASCAP used to tell blacks, no we don't have that type of music. During this time Country and Western music would go to BMI which opened the door from the Nashville area, and for blacks as well, everybody was included. The other larger companies started to see what money could be made from the black community, along with the inclusion of the money from the radio industry and suddenly started giving them attention.

In the past history of black radio broadcasting, the Disc Jockeys could make or break a record. Disc Jockey's of today do not break or make records. Today's records are already proven as making it through their secondary markets.

In 1981 Cisco's a well named club in the Atlanta area, owned by the Williams Brothers closed, and with $70,000 of their earnings a new addition was made to broadcasting, i.e. Coast to Coast radio broadcasting. Coast to Coast began with 25 stations under the sponsorship of Coca Cola, M&M, and Soft Sheen. Upon this sponsorship's support additional stations came under the Coast to Coast title, which made a total of 45 to 50 stations and later on to add another 100 to 125 stations. Coast to Coast played the top twenty records which were logged in the Cash Box magazine listing. This didn't come cheap for Coast to Coast, the amount needed to use publication charts listings was in the neighborhood of $1,200 to $50,000 a year. A widespread magazine such as Billboards' would only cost $1,200 a year to use their sheet, but there still were expenses from any of the publication companies in order to use their sheets.

At the inception of Coast to Coast Atlanta's population was 2 1/2 million, which put Atlanta in the top ten of black listeners. Since it's inception Coast to Coast has grown and is still growing with listeners each time it's broadcasted.

From Coast To Coast To . . .
Inspirations Across America!
Burke Johnson

Burke Johnson, one of the well known Disc Jockey's has been spinning records for years. Burke's job as an air personality permeates from coast to coast with his hosting of Inspirations Across America. Burke has had a successful career as he has enjoyed working with Motown Recording Company, and Jack The Rapper to name a few.

CHAPTER FOURTEEN

IRENE WARE
Irene Johnson Ware

Irene Ware was one of the first black female Disc Jockey's to hit the air waves when she began her career in 1961 on Gospel radio when she hosted The Mandy Show. Tody, she has a lot of experience under her belt; but, still finds time to be very active with her community and has held positions such as: Secretary, Black Music Association, Semper Fidelis Federated Women's Club, Gospel Music Association, Urban League, SCLC, NAACP, and PUSH. Irene received the 1966 Open Mike Magazine Gospel Personality of The Year, BRE's Radio Manager of the Year in 1980, BRE's Woman of the Year in 1977, Jack The Rapper's Roy Hamilton Award winner. MVW of America Announcer of the Year, Citizen of the Year in Prichard, Alabama in 1983, The Original 13

Award at the 1993 Jack The Rapper Family Affair, and for several years she held the position of Southern Region President of the National Black Programmers Coalition, as she remained host of WGOK-AM Gospel program.

LES ANDERSON

One of the disc jockeys that got started in radio on Oct. 11, 1962 at WABQ in Cleveland, Ohio explains how he got the chance to be trained by Jack Gibson and Ed Casslebury. And his starting in radio with those more experienced, he feels that he fit in with their careers like a well nurtured baby he could stand by himself when cut loose. And admits that standing in the control room he had to pick up whatever he could with the various situations going on at the station. He later became very active in the community with the YMCA and the younger kids of the community because of his being a local athlete from his high school years. Les Anderson recalls how Jack Gibson came up with a format for an idea called, 'Tiger Radio', and each time listeners would hear the tiger roar, they called in for a chance to win prizes. Les also recalls that Jack Gibson was the one who taught him how to personify commercials, and Ed Casslebury who would jokingly always come to the mike by saying Ohhhh! and by the time he hit the mike he would be creative with something and make them all look with ahh. Also, he admits imitating Ed Wright who later became a close friend and tells of how they worked together as a team. And remembers being a part of the NARA and NATRA conventions as he got the opportunity to touch base with others who's pictures were the only thing he had to recognize them before meeting them at the convention and later he became a regional president for NATRA and at one time he was also one of the board of directors.

As his career expanded he got the opportunity to break 'Let's Move and Groove' by Johnny Nash, and the song 'Will You Wait' which were a few of the well known song he assisted with introducing to the community. Later, he moved to Houston and for one year he worked at radio station KYOK during the same time that the station was changing management and ownership. And as a result of this change he wasn't happy with this change or living in the southern region. As Les Anderson began to work his way back north, he had one goal in mind which was to own a radio station, and today wonders if the opportunity had been provided could he have been a sportscaster. But, R&B black radio was beginning to change a lot for the disc jockeys and he realized that after he left his final radio station position at WDIA in Memphis it was time for him to do the things he always wanted to do which was to work in the field of Promotions. And after he attained the position of Program Director he saw that he had achieved the learning experience which he had wanted to attain and thus left the field of voice broadcasting and began to work with his promotional career with Warner Bros. Records as Director of Marketing, and today has his own independent, marketing, and promotion company.

In closing out his views of radio as the personality jock era was beginning to end, Les explains that the push of the button became the norm for the industry and therefore the creativity was taken away. And admits that because of this creative form of art being removed, no one cares what time a disc jockey is coming on the air to hear their style of broadcasting as they had in the past. Today, they only want to hear the music played and that's what they're getting.

BERNARD HAYES

Known to his audience as Bernie, he started his career while enlisted in the Air Force in 1950. Afterwards, he received his degree from the University of Illinois in 1956 and shortly after became the first black on-air personality and news announcer in Alexandria, Louisiana. He has since worked for midwest stations ;such as, WMMP, WSBC, WGES and WVON. In 1964 he joined the staff of KSOL in San Francisco. He later moved to St. Louis and was employed at stations KXLW, KATZ, and KWK; which, is where his most negative experience in broadcasting began.

During the spring of 1977 this 22 year career radio veteran began to see the voice of racism attack he and other co-workers. This station which he had come to know and love was also known to the community by repeatingly advertising itself as "stereo in black" recognizing it's dedication to the black community.

Bernie Hayes had been working hard with his daily routine as Music Director during the hours of 10:00 pm to 2:00 am when suddenly, he was changed without notice to a new time period of 2:00 am to 6:00 am. As a result of this and the changes to follow Bernie Hayes was successful in getting the immediate attention from the community and thereby received more than 3,000 petition signatures which called for the removal of Scott St. James from the position he had been recently assigned to. The petition read as follows:

Petition

WHEREAS, Radio Station KKSS came into the St. Louis Market on April 9, 1975 with an appeal to the Black Community for listenership; and

WHEREAS, the recent promotion of Scott St. James to Musical Director of Radio Station KKSS, a position in which he (a white male) will select the kind of music a black audience should hear, and

WHEREAS, very talented and capable Black Radio Announcers, Scotty Lawrence, John Gardner, Bernard Hayes and Johnny Jones, have not been selected for, nor presently maintained in this position by Station Manager Alan Eisenberg, and

WHEREAS, this slap in the face of these Black Professionals and the Black St. Louis Community will not be tolerated.

NOW THEREFORE, the following undersigned residents of the St. Louis Community and owners of business establishments support the removal of Scott St. James as Music Director, and Alan Eisenberg as Radio Station Manager of KKSS; and further support the ownership of this station by persons who will properly recognize and accord the Black Community of St. Louis and who will not make decisions that are slaps in the face of Blacks, to the point of demeaning their character.

We the undersigned have authorized the submission of our names on this Petition to the Federal Communications Commission:

This type of petition filing reflected the emotions of those in the black community who strongly felt that a white man could not dictate to a black audience what type of music they should hear.

As a result of these feelings and the filing of the petition, newspapers in the area were beginning to report headlines of this situation on a daily basis. One newspaper reporter Red Wilford on March 10, 1977 went as far as to say: "THE BIG CHANGE that was recently made at Radio KKSS in their musical programming and personnel has caused me and a great many others to turn our dials to KATZ-AM, a real Black oriented radio station, and I must say that Doug Eason's morning show is 100 per cent more pleasant to listen to than what you'll now find on KKSS-FM. Let us hope that DOUG EASON take advantage of the situation by continuing the good work he has done for so many years. KKSS-FM may have had reasons to cry over their last ratings, however, the next go-around may cause someone to commit suicide, and 'THAT'S A BLACK TRUTH' ... "

Another newspaper reported on the same date the following: "SUIT FILED AGAINST KKSS, by Lily Ann J. Mitchell.

More than 1,000 persons in the St. Louis area have signed a petition urging the immediate resignation of Alan Eisenberg, general station manager of Radio Station KKSS—and numerous employees and former employees of that Black-oriented station have filed a class action bias suit against the station, it was reported at a noonday community-supported press conference on ARGUS press date, Wed., March 9.

Apparently, the past two week's friction at the station came to a head when popular disc jockey, and now newly named Community Affairs Director Bernie Hayes, was reportedly told by Eisenberg 'to take a two week vacation.'

'I was not told I was fired. It really surprised me when I was told to take a two week vacation because this thing was hurting me and the station. The station manager, Eisenberg, told me it was not a suspension, but a vacation until things were quieted in the Black community,' Hayes stated."

And as the plot thickens, on Thursday, March 17, 1977 it was reported in the local newspaper, "KKSS MANAGER "Blows" It; Fires Scottie Lawrence, Helen Hagen"

"In an interview, Hayes said that Johnnie Jones, who replaced Hayes as program director recently, told him he was dismissed Friday. But Alan Eisenberg said "that was a misunderstanding" and that a final decision on Haye's status would be announced at a press conference Friday, March 18.

Eisenberg would not state the reasons for Lawrence's dismissal saying the former disc jockey would be informed in a "service letter".

The station manager denied that Hagan had been fired. 'As far as I'm concerned, she resigned.' Eisenberg said, she told me she was financially unprepared to move to St. Louis and that she was going back to Philadelphia.

Hagen told the American that Eisenberg failed to follow through on a verbal agreement to assist her in finding housing in St. Louis and transportation to and from the station at 1215 Cole Street."

And Hagen responded with the fact that Eisenberg informed her that she had aligned herself with Bernie and that she would not have a job come Monday.

When she reported to work on Friday night, program director Johnnie Jones asked for her keys. Hagen recalled looking up on the wall of the station where licenses were placed and realized that her license along with Bernie and Scottie Lawrence's were missing.

But, there were those in the industry and community who also recall statements by Eisenberg who bragged about the false fact that his station was the first to have a black female disc jockey. As a matter of fact, he's twenty years late, previous black female disc jockeys were Willa Mae Gracey for KATZ, Yvonne Daniel at WATM which was formerly called WTMV, who's also the daughter of singer Billy Daniel, and more recently was Rita Hunt at WESL.

And then there were others who felt that Scott St. James was being groomed for the top spot of manager at the station with the understanding that Alan Eisenberg would be moved to a warmer climate in Texas. Of course this was denied by both.

But what wasn't denied was the reason why all of this happened. Eisenberg explained to everyone that this was not an act of racism. Eisenberg was stated as saying, to the St. Louis Sentinel on March 17,

1977 the following: "I see no problem with Mr. James in that position, when you have Mr. Andrew Young U.S. Ambassador to the United Nations, representing the United States, when America is predominately white."

And another finding was noted along with the copies of the paperwork which detailed a $125 a week wage increase for St. James.

And as a result of all of these findings a committee was made which was comprised of representative of all facets of St. Louis black groups, conservatives, moderates and activists. From groups such as the NAACP, Urban League, License Collector Benjamin L. Goins, the Rev. James Cummings, School Board Members, State Senator J.B. Banks, ACTION chairperson Percy Green, and Louis Omar of the World Community of Isam in the West (Muslims).

The spokesperson for the group, Chris Moore of TV Channel 9, was critical of the station's policies and reportedly informed the stations owner, Joseph Amaturo, who had flown in from Florida for the meeting that there is a lack of community involvement (especially toward blacks) and there was the question of difference in salaries due to race. The fact was noted that Hayes, a 22 year veteran of radio and one of the most experienced and articulate disc jockeys in the area was making less than his subordinated and only white disc jockey at the station.

It was also stated by Morris Henderson, the sports editor of the St. Louis Anerican, that he will not participate in anymore of the stations' talk shows.

And while all of this was becoming quite apparent to the owners as to the impact of their bad management decision, the community was beginning to show their support by asking advertisers to discontinue using KKSS radio. The community was also responsible in assisting in sending a petition to the Federal Communication Commission, complaining about the discriminatory practices.

The previous actions taken were not enough for those involved and on March 17, 1977 the ARGUS newspaper headline read: "$300,000 Suit Filed Against KKSS".
"Three dissatisfied off-duty KKSS-FM employees filed separate suits against the Black-oriented FM station, totaling some $300,000.

Bernie Hayes, Scotty Lawrence, and Helen Hagan filed an Equal Employment Opportunity (EEOC) discrimination suit as well as a personal damage suit against Alan Eisenberg, station manager and Joseph Amaturo, president of the Amaturo group, and owners of the station as result of being told individually that they had been fired.

It was reportedly said by Eisenberg, that this is a "tremendous over-reaction to the whole situation by the Black community over a few personnel changes.

And while these changes may have seemed only a small view point from the thoughts of Eisenberg, the following finding displayed just how big of a deal this really was as the case was quietly settled out of court.

Today, Bernie Hayes enjoys working at KTZ doing a morning talk show, and he now owns his own record company, and a beauty salon. Scotty Lawrence is now a pharmacist. Helen went back to Philadelphia and decided to drop out of the suit. Eisenberg moved to Denver, Colorado a few years later and is working with a cable company. Johnny Jones moved to California, and John Gardner is working with KEZK radio station.

Rufus P. Turner

The community awareness for radio was not only viewed from a social, entertaining and economical side but also from the scientific level as well. There were others who were also enthusiastic about the power of radio on a scientific level. On October 27, 1934, Rufus P. Turner was recognized by Robert Ripley of Ripley's Believe It or Not Discoveries.

Rufus P. Turner was highlighted as the builder of the tiniest radio set, and was called a "midget radio expert". After rushing from his lavoratory in Waltham while breaking most of the speed records, Turner soon came upon Mr. Ripley at the Hotel Touraine in the downtown Boston area on October 22nd right after a squadron of motorcycle policemen had escorted Mr. Ripley to the hotel. Although he had been warned that he may not be seen by Mr. Ripley, Turner remained true to his fate and upon meeting him he found him to be warm and courteous.

As the two of them were surrounded by numerous reporters, Turner's face gleamed with joy as he displayed his midget radio. First, he proved that it could be passed through the eye of a needle, and, afterwards Mr. Ripley asked him if he could have it. The newsmen also begged him for it, but, he decided to keep it for his mother as a souvenir.

Mr. Ripley promised to feature his midget radio set in his November cartoons and as he read through some of the clippings in Turner's scrapbook he came upon one of his very own drawings depicting Turner's first midget set which was built on a straight pin and that it was also in the Ripley show at the World's Fair. As a result, Ripley autographed the clipping and one of his "Believe It or Not" Books.

Ripley also revealed through the meeting that he was born on Christmas Day and found out that Turner was too.

And even though Rufus P. Turner may not have been into the vocal side of broadcasting he was truly aware of the remarkable discoveries the technology could produce for radio announcers.

Another concentration from the community was the power of broadcasting bases from your very own home. There were those who knew what it could mean if they worked hard enough to get an FCC license and what power they could have if they could operate from their own home base. Their recognition can be included in that of the black radio history because of the following young radio operator.

Before this recognition had been given there had only been recordings of achievements made by men as radio operators of stations; but, in April of 1935, Miss. C. Geneva Lyman received her license as the first female Negro to be granted an amateur radio license. Although, Miss. Lyman may not have been a Disc Jockey, her perseverance to perform such an endeavor during these times was a milestone in itself.

"IF THEY HAD KNOWN..."
HIGHLIGHTS OF DISC JOCKEYS BY HINCKLEY

Because no one has ever taken the effort to project what this generation of Disc Jockeys did for their community, and careers, some of them got together in New York and interviewed with USA Today. This was an attempt on their part to tell their story and give them a place in history.

"If They'd Had Any Idea How Big This Was Going To Get, They Never Would Have Allowed It"

The Rhythm, The Blues, The Black Jocks of New York
(They tell their story)

Back in the '60's, a rookie disc jockey named Bobby Jay rode the subway to work every day at WWRL. Sometimes, he rode with an older man named Tommy Smalls, a record promoter who every Wednesday would bring his company's new releases to program director's and attempt to persuade them that these songs should be added to their stations' play lists.

R&B stations and their jocks had always been suspect outsiders, a point underscored during the 1965 Watts riots, when black Los Angeles DJ Nathaniel "Magnificent" Montague (who earlier worked at WWRL) found himself called an instigator because some riot participants appropriated the-on-air slogan, "Burn Baby Burn". Montague was really talking about hot tunes, not torching the neighborhood.

But it wasn't until 1971, when Manhattan Borough President Percy Sutton's Inner City Broadcasting Corp. bought New York's WLIB and WLIB-FM, soon to be renamed WBLS, that "black" stations became generally accepted in the mainstream of urban markets.

Not that fear has totally evaporated; as recently as a couple of years ago, police were taping New York's WLIB for signs of sedition. Generally though, the R&B-and soul-based formats nurtured by WBLS are as accepted today as Top 40 or easy listening. So accepted that one of WBLS' most successful programs, "Sunday Morning Classics," features many of the artists Smalls used to play: the Spaniels, Jackie Wilson, etc.

The host of "Sunday Morning Classics," is Hal Jackson, a pioneer black DJ. An upbeat jock whose radio patter runs to lines like "Think that you can and you will," Jackson just kept thinking positive. Your listeners don't want to hear about your problems," Jackson explains today from his cozy, memorabilia-filled office at Inner City. "You're an entertainer. When the red light goes on, it's `Hello Hal, Everybody's Pal.'"

Last fall, Jackson celebrated his 50th year in broadcasting-a satisfying achievement for a man who had once been told, when he applied for his first professional radio job in Washington in 1939: by a station he wanted to broadcast at: "No nigger will ever broadcast on this station."

THE BLACK DISC JOCKEY DID NOT enter American culture like James Meredith enrolling at the University of Mississippi. The early black men, and few women who were allowed to penetrate a white institution were only allowed because they were considered useful, harmless or both.

In early 1947, Ebony magazine counted 16 blacks among an estimated 3,000 disc jockeys in the U.S. . But popular music was changing, and one of the major changes was the rise of black music, from the pop/jazz/R&B vocals of Ella Fitzgerald, Billie Holiday, Ruth Brown, and Dinah Washington to the swinging rhythm and blues of Paul Williams, Louis Jordan and Bull Moose Jackson and to the smooth sounds of the Orioles, the Ravens and Charles Brown.

Furthermore, this new music was not reaching blacks alone. Millions of whites, hungry for an antidote to Guy Lombardo and Frankie Laine, also were finding it-frequently through black radio and black disc jockeys.

"Playing black music on a wide scale started down South, "says Ed Casslebury, who got his first DJ job in Birmingham in 1950 and later worked for the New York-based National Black Network, which held syndicated programming to more than 150 black stations." The white and black kids there all loved R&B. When we'd have shows for the black kids, the white kids would sit in the balcony also, of course, it was segregated."

Playing black music, then, was a shrewd move-sort of. "Remember, black music was subversive then," Jay says. "It was going to ruin America."

Fortunately, capitalism intervened. "Companies were trying to find ways to reach the black market then," says Evelyn Robinson, who from 1952 to 1959 co-hosted "Life Begins at Midnight" from Harlem's Palm Cafe every night over WOV. They were beginning to realize blacks used Hoover vacuum cleaners and drank Coca-Cola. Black radio could sell listeners those products."

Or just make them tune in at all. In 1946 there were 943 stations in the U.S. by 1951, there were 2,000, and competition was tough. "After the war, a lot of stations were hurting," says Jay. "So some white-owned stations started programming black music—because they'd tried everything else and they had nothing to lose. Then, it took off."

Today, there are close to 7,000 stations in the country, hundreds of them black." If they'd had any idea how big this was going to get," says Jack 'The Rapper' Gibson, a pioneer black DJ and publisher of an influential trade sheet, "they never would have allowed it." He chuckles.

IN NEW YORK, TOO, THE BLACK DJ entered quietly. When Newarks' tiny WHBI decided to try black music after the war, it first turned to Danny Stiles, a white DJ regarded as a little eccentric. "Actually, I prefer avant garde," laugh Stiles, who today broadcasts on WEVD. "I just never thought of music as black or white, I'd play Billy Eckstine, Billie Holiday.

"There were a few jazz shows then, but it was very unusual for anyone to play black music. White DJ's almost never played it, and black DJ's hardly existed, at least in the North. In the South, there were more. But I didn't run into a black DJ until I went to WNJR several years later."

There had been some black programming on radio for years-but the truth is, the best known "blacks" on American radio at this point were Amos `n'Andy.

Eddie O'Jay, later one of New York's great personality jocks, discovered he was building from scratch. "First the station had hardly any records my listeners would want. It was a black station, but the library was all Patti Page, big bands. None of what were called `race' records. So I'd drive to Chicago to stock up. Once it became clear my listeners loved it, so did my station."

By 1953, Variety estimates 25% of U.S. stations carried at least one rhythm and blues program. Yet the field didn't become saturated, because as late as 1956 big "pop" stations still were petrified of black music, as reflected in their desperate affection for bland white "covers" of R&B tunes.

"The kids knew the real thing." says Douglas "Jocko" Henderson, the Rocket Ship man and one of New York's most powerful black Djs over WLIB, WADO (the former WOV) and later WWRL. "When I held my first Rocket Ship show downtown on Broadway, the audience was 85% white. I usually used the Apollo, but some fans wouldn't go there."

"When pop radio finally realized it would have to change." Casslebury says, "that's when they started calling this music rock `n'roll. They needed a way not to say rhythm and blues; even though it was the same music."

"When I got to New York in 1953, there was a hunger for R&B," says Hal Jackson. "The people who played it became heroes."

Other jocks echoed that thought, which underscores an irony: The multiracial popularity of the music, the stations and the jocks as institutions in the black community.

"Those were golden years for Harlem and the disc jockeys helped bring it together." says Evelyn Robinson, Sister of Sugar Ray and one of the first black women Djs, she hosted celebrities and friends at the Palm while playing music from Duke Ellington to Elvis to Ruth Brown.

"I used to walk to work," she says, "and all along the way, people would all stop and say hi. I'd have Nat King Cole or my brother on the show and people would come by to talk and keep up on what was happening. Some nights, I'd feel great and somebody would call and say, `What's the matter, Evelyn? You don't sound like yourself tonight." People paid that kind of attention because the disc jockeys were your neighbors and friends." And there were quire a number. Willie Bryant, The Mayor of Harlem, Dancer, actor, singer. He broadcast from the front window of the Baby Grand on W. 125th Street, often calling out the names of people who passed on the street. Hundreds in a show. When rock`n'roll became big business, Bryant, criticized stations that hired whites to play it. We made this music popular, he argued. Now you're cashing in.

Ralph Cooper, the Black Clark Gable on film and originator of "Amateur Night at the Apollo," did another midnight remote from the Palm Cafe, over WHOM (and before that, WOV and WINS).

Hal Jackson broadcast live from Birdland as well as from the WLIB studios on the top floor of the Hotel Theresa. Jackson also landed a rare movie role, as host of the R&B film "Rockin' The Blues." Jack Walker, "The Pear Shaped Talker," had several shows on WOV. He also became program director at WLIB when he was fatally stabbed on New Year's Eve 1970 by a probationary engineer he had dismissed. He was 48.

Jocko in addition to the radio, had an afternoon TV show on Channel 13 (WNTA). Herman Amis broadcast live from the Palm Cafe, over WOV, Amis now at WNJR, whose other Djs included the flamboyant Ramon Bruce ("I am the Bruce called Ramon!").

Smalls who actually was the second Dr. Jive (after Phil Gordon), had the kind of on-air cool all female listeners wanted. There was Joe Bostic of WPNX, Rocky G. George Hudson at WNJR, Carlton "King" Coleman at WWRL. Not to mention two national legends: Magnificent Montague and Maurice "Hot Rod" Hulbert, an early rhymer from whom Jocko learned some of his style. O'Jay was at WLIB from 1965 until he left to become deputy commissioner of sanitation in the Abe Beame administration. Today, he's back in the broadcasting industry.

It's also worth remembering that while many of these jocks were educated people, what they did on the radio was largely self taught.

"It wasn't like today, where you're trained for the radio." says O'Jay. "The career schools' wouldn't take blacks anyway. I got my start by going to a local station and saying I wanted a job." On his second try, he got one.

"I was asked to go on the radio because they wanted a woman's voice to help sell products," recalls Evelyn Robinson. "I had no experience, but I took to it naturally. The main thing, I found, was never allow any dead air. At first I was on with Jack Walker and I didn't talk much, but that changed. They never let me pick records, like the men did, but by 1957 and 1958, I did my own shows at the Apollo."

"In fact picking records was a haphazard affair. "Not many people," says Jocko, "really knew how to program." So, they compensated with style-and the only rule was to keep it lively. No dead air.

Jocko, the son of Baltimore's first black superintendent of schools, spoke in rhyme ("Back on the scene/With the record machine/Correct time/Is 10:19") with the diction of a Shakespearean actor. "I always saw myself as an educator," says Jocko, who today lives in Philadelphia and develops school learning programs. "I'm in the same business today I was 25 years ago. I'm just not on the radio."

Casslebury, on the other hand, remembers that a serious, articulate approach earned him a flat turndown at his first DJ tryout.

"I auditioned clean. When I was done, the owner said, `We can't use this guy. He talks too good. He's not splitting' no verbs, cracking' no adjectives.' He wanted screamers and shouters. A lot of owners were like that. They wanted what they thought black listeners wanted."

One reason Bryant left his Baby Grand show was his refusal to read ads he considered derogatory. Frankie Crocker, a jock famous for flamboyance himself, finally threw down the gauntlet when he became program director of WBLS for Inner City in 1971. We will take no ads, he said, that insult our listeners.

'One Other Concern Of Early R&B disc jockeys was the fact that they weren't making enough to live on.

"The pay was horrible," says O'Jay. "Terrible," agrees Casslebury.

White jocks weren't cleaning up, either, but black programming was particularly notorious for low-budget owners keeping expenses even lower. Accordingly, Djs supplemented their income through shows.

Thus Jocko had daily radio shows in both New York and Philadelphia, plus side projects often reserved for white Djs: the TV show, compilation record albums, record publishing.

Hal Jackson once juggled four shows. "I'd do Washington in the morning, then drive to Annapolis, then Baltimore, then back to Washington at night. I did this for three, four years. I don't know how."

Chicago's Al Benson started his own record label. Eddie O'Jay for a time owned the group the O'Jays. But bread-and-butter for most Djs was live shows.

"We felt like gods, "says Casslebury. "People would call us with their problems-like, my sister's having a bad time, could you talk to her. And we'd do it.

"They trusted us because we were out there."

"The key was always to keep up with what your listener wanted. Every night, I'd be out hearing what the people were saying, listening for new phrases, new music. Then, when I was on the radio the next day, I could say, "Got a request from Jack. Saw Jack last night'. It gave me an edge. I've been off the air 15 years now and there still isn't a day goes by that somebody doesn't recognize me on the street, or at a restaurant."

"In the 50's, these guys were role models," says Bobby Jay. "The way they dressed, the way they talked. It's not like the Superfly thing. Their whole lifestyle just looked good.

SO WHO GIVES THE EARLY JOCKS the credit they deserve?

Artist? Some of them. "In the 50's, artists couldn't stay in southern hotels," Casslebury says. "We'd put 'em up, have 'em on the show, play their records, book 'em and feed 'em. We took care of each other.'

Record Companies? No. In 1946, recorded music was a $150 million industry. By 1959, it was up to $650 million, much of that R&B and rock 'n' roll. When Berry Gordy, Jr. started Motown, he based his strategy on black jocks. "He'd offer us the Supremes for nothing," recalls Gibson. "Just to get exposure. He'd drive them down himself."

Fans? Rarely. Had rhythm and blues not forced its way into mainstream America popular music of the past 40 years might have taken a far less interesting course.

"You could see by 1952 that this was where the music was going," says Danny Stiles. "You could tell black and white kids were all crazy for rhythm and blues. We got 5,500 people at a dance in Newark. As a disc jockey, it seemed clear to me this was the place to be. I just went along for the ride."

"Artists like Fats Domino loved Freed," says Hal Jackson. "They realized he did things for them that we couldn't have done.

That is to say, even though Jocko had a better TV show than "Bandstand," no black show was game to get a national TV deal. Black jocks couldn't get the record and artist tie ins that went to whites like WINS' Murray the K (The Fifth Beatle), and only a few Jackson, Crocker, Chuck Leonard) got hired by the more lucrative "pop" stations.

"Without personality," says O'Jay, "radio becomes a jukebox. Anyone can push buttons. It's been my belief that formatting is a method to control the possible power and persuasion of the jock, by limiting his input on the air."

Casslebury agrees: "They deliberately took away the DJ's personality. It was a way of cutting his power and keeping more control."

Whatever the intent, pioneer black jocks have largely become invisible. They are cited here and there-Wes Smith's new book "Pied Pipers of Rock `n' Roll" (Longstreet Press) makes the effort-but in most rock`n'roll history, black Djs are mentioned less than Pat Boone.

Alan Freed was a charter inductee at the Rock and Roll Hall of Fame; the Hall has no black Djs except B.B. King, who broadcast over WDIA in Memphis). The National Association of Broadcasters last month tapped Hal Jackson for its Hall of Fame-the first black so honored.

For all these reasons, Jack Gibson last fall started the Black Radio Hall of Fame, on the Atlanta street where Martin Luther King, Jr. is buried. The charter class, including two whites, was Oscar Alexander (Winston-Salem), Benson; William Brown (Beaumont, Tex.,) Ed Cook (Miami and Chicago); Merri Dee (Pittsburgh); Dave Dixon (St. Louis); Gibson; Jocko; Joltin' Joe Howard (Detroit); Al Jefferson (Baltimore); "Chatty Hattie" Leaper (Charlotte); George "Hound Dog" Lorenz (Buffalo); Larry McKinley (New Orleans); O'Jay; John R.; Rudy "The Deuce" Rutherford (Columbus, Ga.); Milton "Butterball" Smith (Miami); Walker; and Bill Williams (Detroit), and the roster goes on and on.

"Is this an honor? Oh, my God, yes," says Eddie O'Jay. "I know it's a cliche, but it isn't all the time in my age bracket when you get to smell the roses while you're still here."
Or sometimes, in his profession, when you get to smell them at all.

To show the improvements taking place with the likes of disc jockeys like the ones interviewed above we can find some of the necessary information in the south, where changes were made when Winston-Salem, North Carolina began it's first 'Negro Radio Station'.

Payola

Even though I've had much hesitance in writing on the subject of Payola, I have been asked by many of the disc jockeys to include it because it was also a part of their broadcasting history. And as I searched through the pages of stories written on the subject I have decided not to name those infractured by such a negative undertaking developed in the industry by those who desired to attack the professional abilities of the disc jockey and turn the many years of triumphant achievement into a career of undermined greed.

The following information written on the details of the Payola trial can be found as an interest for those studying the field of broadcasting, who without reading this information may not have known that certain practices can cause infractions for those wishing a career in broadcasting, otherwise, it wouldn't be of any interest other than gossip.

To begin to understand how Payola is included into the industry of broadcasting, we must understand the definition of the word 'Payola'. According to Webster, the following definition of Payola was given in the late 60's. "pay.ola: undercover or indirect payment (as to a disc jockey) for a commercial favor (as plugging a record)".

As a result of this type of process existing in the career of the disc jockey, a few were charged with practicing the policy of payola and were highlighted in the 1964 New York Post before all charges had been proven to be true or not.

The first arrests consisted of five disc jockeys and two radio station officials. Their ages ranged from 25 to 40 years old. There was to be one more arrest at the time of the printing but the individuals' name had not been listed and they had not yet surrendered.

The ones who were arrested were charged with commercial bribery for the acceptance of payola. It was also mentioned that because of their styles of playing a record and singlehandedly making it a hit was the reasons why record companies would attempt to confront these particular disc jockeys. It was also said that when they played a record many times in one night that they were plugging it because they had been paid to do so, (unfortunately, this had to be proven), we must also think of other reasons why songs could be played over and over during this time. There were times when the request lines asked to hear it again. Other times it was the song of the day as requested by listeners who may have heard about it through friends and wanted to hear it for themselves. The phenomenon of rhythm and blues was still taking the community by storm in 1964, and if a listener wanted to hear a song again most disc jockeys would oblige and play it again for them. This is not to say that payola wasn't a part of the playing statue, but, it is to let those who may think that what might have been sincerity as a disc jockey doesn't necessarily mean that they had any other endeavor in mind.

As the arrest were made there were many more to come. The public was being informed of the infractions of what would happen if someone was found guilty of accepting payola during this time and if this was found to be true of any one charge against a particular disc jockey then this disc jockey would have to pay $500 for each count and one year in the city penitentiary as a maximum penalty.

The amounts of the payola charges to the disc jockeys ranged from their taking anywhere from $7,420 to $36,050 from various record companies, and distributors over a period of time. After pleading not guilty to the charges all of the disc jockeys were paroled for their individual trials.

As the grand jury gathered the facts of each individual charged their lives were still being haunted with the monstrous charges placed on their careers. There were suspensions, firings, and demotions at many of the stations as each station manager wanted to look impressive to their listening audience, and also because they were afraid to stand up before the grand jury and support their people. Unfortunately, the disc jockey was charged as guilty before the trials began and their careers were being damaged by the media during each printing.

It took six months for the grand jury to complete the investigation and the findings were as follows:

After questioning seventy persons in the music industry, and reviewing the books of about 100 companies which were all subpoenaed and examined, they determined that there were three principal methods of payola uncovered: direct cash payments, regular weekly or monthly payments, and royalty payments. There was mention that of the 23 record companies and distributors involved in this case that there were five big spenders, three of which were distributors and two record companies which had payout totals of $65,202.00.

According to sources interviewed only one black disc jockey reportedly went to jail for a short period of time for accepting payola, who by the way was none of the original ones charged, and as result of his endurance of the trial and later retiring from the career of broadcasting in his later years, the rest of the story is his to tell as he completes his manuscript.

At this point it up to those of you who have read the above information concerning payola to decide what this information means to you as you review the career of the disc jockey and as for me it means nothing more than to do what's right in the business and make sure that you have covered your tracks with any relationships in the industry.

Open Mike
NARA Magazine Excerpt—Dec. 1964

DeeJay of the Week—Kae Williams-WDAS Philadelphia, Pa. written by Frankee Davenport.

When most veteran broadcasters congregate and the discussion of the top disc jockeys in Pennsylvania, the names of Kae Williams, Raymond Bruce, and Randy Dixon. Always mentioned as pioneers. Now Ramon is gone, Randy is active in only newspaper circles, but Kae is still the broadcasting pace setter he was 20 years ago. Son and grandson of a South Carolina Minister, Kae originally started out as a song and dance man, but, when this became physically impossible the instructor said the Dolphin school of dramatic arts directed advised the bass voice Kae to go into radio. He did exactly that, and the rest is broadcasting history. Kye's novice stations were WPWA, and WIBG where he is most widely known for his past office affiliation at WHAT and affiliation at WDAS of Philadelphia, Penn. full of vye, Kye always seems to be in thick of controversial issue which later breaks into newsprint. Such as his stand against jazzing up spiritual music and camping against suggestive lyrics and sex latent disc. Both of these received national attention, but, much of this can be traced back to Kye's christian upbringing, but, some is his straight forward way of telling it like it is to anybody about anything. Good example is Kye's 90 day report of NARA which is published in Open Mike and fully endorsed by us. In addition to Kye's tremendous force for good, he has also given great talents to the record industry, with recognition of great talents such as Solomon Burke, The Cavaliers, Jon Thomas, The Sensations, Rollie McGill, Lou Andrews, Yvonne Baker and the Sensations, Screaming Jay Hawkins, Bobby Bennett, Doc Bagby, and the Silhouettes. In 1958 the Silhouettes get a job, received a gold record and today Yvonne Baker of the Sensations has a monster in "What a difference love makes", as does Jon Thomas in "Feeling Good". Most of the artists discovered by Kye have been personally managed by him and recorded on Kye's label, Junion record company, named after his son Kye Williams, Jr.

During this time headlines were made of Sam Cooke being killed by an irate woman, and motel manager in Louisiana.

RADIO RACE RELATIONS INFORMATION CENTER
Radio Race Relations Information Center

In 1970, the Radio/Race Relations Information Center reported their findings and they reported as follows:

Radio/Race Relations Information Center (1970)

These findings of the Race Relations Information Center of Nashville are a detailed study of black-aimed radio programs entitled "How Soulful is `Soul' Radio?"

The center, a private non-profit organization that collects and distributes information about race relations in this country, is headed by Luther Foster, president of the Tuskegee (Ala.) Institute. John Seigenthaler, editor of the Nashville Tennessean, is vice chairman.

"Over-all," the report states, " `soul' radio's responsiveness to the black community showed a marked increase in the sixties, with the greater changes occurring in public affairs, advertising, news broadcasting and equal opportunity. But few broadcasters showed any willingness to move until prodded by black sentiments (and new Federal Communications Commission vigilance) and some still have moved only very slowly." The report says it can be expected that pressure on broadcasters from the black community will increase "forcing either a boom in black radio entrepreneurship or radical changes in white broadcaster's policies."

A recent development has been the formation of the only black-owned radio chain in the country, the result of three station purchases by singer James Brown. Mr. Brown purchased WRDW in Augusta, Ga., WJBE in Knoxville, Tenn., and WEBB in Baltimore. Each of these stations is well-rated both in studies of black audiences and in general station surveys.

The report lists Mr. Brown's three stations and only six others as owned by blacks among the 310 stations in the country that program for black audiences. These are KPRS in Kansas City, Mo., KWK in St. Louis; WCHB and WCHD-FM, in Raleigh, N.C. Other estimates of black-owned stations have been higher. Last month, the magazine Advertising Age listed the same stations as those cited by the center, but added WMPP in Chicago; WEUP in Huntsville, Ala.; WTLC in Indianapolis; WOR-TV in Hattisburg, Miss.; WWWS-FM in Saginaw, Mich., and WVOE in Chadburn, N.C.

The center studied the programming and hiring practices of five white-owned black-oriented broadcasting chains and concluded that there appeared to be some instances of blacks being given titles but no responsibilities.

The five chains have 22 stations, including Rollins, Inc. Broadcasting Division, and Rounsaville Radio Stations in Atlanta; the Sonderling Broadcasting Corporation in Nashville; Speidel Broadcasters, Inc., in Columbia, S.C., and the United Broadcasting Company in Washington.

The survey found that hiring practices at Rounsaville and Sonderling were the fairest of the chains, and added: "Even some critics who argued that no broadcaster can take pride in his personal record acknowledged in the achievement of these two firms."

Each of these stations relies heavily on contemporary, popular rock music and rhythm and blues. Each one says it broadcasts this music in response to listeners' demands, but, critics, the report states, dispute this. William Wright, director of Unity House in Washington, says "the problem comes with white broadcasters who have brainwashed black people into accepting 24 hours of `soul.' They've created a `soul music' mentality, and black kids are paying for it."

Nearly a quarter a century after a radio station first geared its entire broadcasting format to black interests, there still isn't a nationwide black-oriented news network," the report says. "Blacks still comprise the vast minority in key executive stations at `soul' stations. . . . All this troubles black radio reformers. In addition, they wonder if white management's public affairs efforts are meant to enlighten and serve the masses or merely to satisfy minimum F.C.C. requirements and remain in business."

And while the study given above was being reviewed radio station WRVR FM outlet of the Riverside Church by the Television, Radio, and the Film Commission of the United Methodist Church along with the National Council of Churches, the National Catholic Office of Radio and Television and the National Urban Coalition began to establish something the community was longing for with a program entitled, 'Night Call', which was designed to allow callers from the Nation's ghettos to call in and speak with their prominent guest speaker. As the community began to learn of this program they were informed that they could call in and speak with the guest speaker during the hours of 11:30 pm to 12:30 am to discuss the various topics. As the station tried to assist those from the communities from having to pay for long distance calls, the telephone company objected because they feared that too many lines would be tied up, as a result, the station developed plans on how to reimburse each caller. The 'Night Call' host was pioneer disc jockey Dale Shields from radio station WLIB in New York. The first guest for the show would be the Rev. Ralph Abernathy, acting Director of the Southern Christian Leadership Conference, and others such as Stokely Carmichael, author of 'Fail Safe', Harvery Wheeler and others.

If other stations would decide to do something of this caliber then there wouldn't be any need to do studies such as the one above.

AWAY FROM THE BLUES
Away From The Blues

1-18-54
Newsweek Magazine

"Most of the 270 Negro radio stations in the country just play blues, pops, and spirituals. But the Negro also wants network quality shows of his own." So says Leonard Evans, who, on Jan. 18, will inaugurate his National Negro Network, Inc., the first attempt to program for colored people coast to coast.

Evans, a Negro who has studied his race's consumer market for twenty of his 39 years, feels sure that he has a money maker on his hands. The network, which will reach some 40 cities with large colored populations, will have a potential audience of 12,000,000 of the 16,000,000 Negroes in America (who earn $15,000,000,000 a year). Evans casts an optimist's eye at his sponsorship prospects.

He has already signed two bank rollers—Philip Morris and Pet Milk—for one of his four soap operas. The Story of Ruby Valentine, which will star Juanita Hall, the original Bloody Mary of "South Pacific,"

And last week he had high hopes of selling two more, Cathy Stewart (with Hilda Simms, who played Anna Lucasta on Broadway) and Cousin Honey, which may feature Ethel Waters. Among the other projects for this year are five half hour shows, including Cab Calloway in It's a Mystery, Man. And we want to develop our own Edward R. Murrow."

NNN, which Evans is forming with cooperation from jack Wyatt and Reggie Schuebel, New York consultants, is strictly a business proposition: "This is not a crusade for the intermingling of the races," says Evans. "We're out to move tonnage, to sell merchandise. If we do help race relations, it's incidental.

Ear Appeal

Neither Amos 'n' Andy nor Beulah would qualify for NNN, according to Evans. First of all, "the Negro may enjoy these shows, but he is also embarrassed by them." And "I can go through their scripts and point out things that a Negro would not say." To keep the dialogue authentic, Evans will hire one colored writer for each show "because he knows the people" and one white scripter because "they're the only ones that have any experience." Evans himself will go over scripts and commercials to see that they appeal to the Negro ear "the same way Negro music does."

"You wouldn't get Guy Lombardo playing in the Savoy Ballroom," he explains, "You'd get Duke Ellington."

MITCH FAULKNER

Disc Jockey and Radio Producer Mitch Faulkner began his interest in radio at a young age when he would listen to disc jockeys from the past as a child and found it to be entertaining. During this time he would use a eight track recorder and make tapes of his friends as he recorded his voice on the tape, as a result, they told him that he should be in radio because he sounded like a deejay. Mitch and his brother soon after began to start their own studio which would record gospel choirs, etc. to make music. As he and his brother began to develop things more and more for those in their area a lady from the local NAACP came to them and asked them to make a tape and as a result they received air play of this recording. Soon after Mitch was asked to come to the local station and make an audition tape for them since it was different than what they had on the air, and as they set approval for his work, he quickly began to get familiar with those at the station, and how the business operated, and as for him the rest is history as he set his career path from this time fourth in 1979.

Mitch Faulkner considers his past in radio even more exciting as he began to meet pioneers of radio from his childhood. As he recalls how he first met Hos Allen, who's station WLAC from Nashville played black music late at night in his area. As he recalls at one of the Jack The Rapper's conventions Hos Allen was getting an award and he remembers how excited he was in getting a chance to finally meet him and to his surprise he found that Hos Allen was white and also handicapped. Mitch recalls that the audience was so excited to honor Hos Allen that they had to stand and look over the applauding crowd in order to get a chance to see him. Mitch also recalls getting the opportunity to meet other pioneers in the early stages of his career when he met Randy with Randy's record shop, Spider Harrison, Bill Williams, Jack Gibson, Herb Kent and others and found this to be exciting to finally get a chance to see the ones he heard during his childhood or had heard about, but, never knew what they looked like.

Mitch believes that the past traits of the pioneer disc jockeys has worked for him because he believes that studying their established traits gives the disc jockey a measure of how to develop relationships with the audience which they're trying to appeal to. As he looks at the way the pioneers from the past broadcasted from their stations versus the current trend of format radio he remembers how pioneer disc jockeys used their personality to change their music style or presentation of the music which they played as each one would come on the air and change the atmosphere with their presentation and their music style. And with this change in the element of style the disc jockey of today who may not have control over the music which is played can be more creative by the means of inserting his personality, being relatable with what the audience, knowing what's current with them, knowing what's happening in the community be it good, bad, slang, or hip, and find a way to present this with the music you're presenting in a way inwhich the audience understands it; the disc jockey who uses these past practices can achieve the personality permeation for their audiences today, along with promoting the maintenance of some of the past traits of the pioneers.

Mitch believes that if the disc jockey of today learns more about themselves and about their community rather than trying so hard to enunciate then they can use their knowledge of the community to project a personality edge which the disc jockeys of the past used. Because if they don't involve the personality trait with the current age of the satellite there will be disc jockeys who won't survive as only a handful of disc jockeys will be broadcasting within the communities of the nation from as far as California to the communities of New York through this method of broadcasting. This difference in the mode of communication will still need you to be relatable and draw upon that personality edge that the disc jockeys from the 20's, 30's and 40's used, but, as today's radio has been found to be reactionary, it follows, and sometimes it's prohibited today's disc jockeys can still go back to what's called the teenage stage when it communicated to people, entertained, informed and led the way.

Along with this the disc jockey of today and the future must realize that change is good in any situation but we all must understand that we can keep the standards in tack through the use of maintaining personality while changing with the times. And while there may be emotional differences with these changes there is a way to overcome the emotions involved by keeping the personality and learning that disc jockeys can be the owners of the stations which they have worked at and thereby control some of the issues of concern.

Another vision for the black disc jockey is to own the station and look at the bare bone success. One example Mitch uses is how Motown became a success by making black music, and they made it so good until it spread and became pop music. He explains that they didn't go into the studio and say let's make pop record with the Temptation, or the Supremes, they went into the studio to make a black record, and the music was so good that the masses accepted it. The same thing can happen with radio, from an ownership stand point, if a person takes a radio station and makes it the best black product they can possibly make it, it will become popular with the masses then from this point it becomes a pop radio station which is the ideal situation from an ownership stand point and it's going to take someone who's financially stable and able to handle the situation to weather the storm. And thereby, take a chance to be different, to be black and to be proud of it and broadcast that fact, but, the possibilities of this happening are very low because people believe that others will resist what they do. Which is also similar to what happened in the 60's and to Rap Music, where music is so good that the masses accept it, black and white, and they don't see it as color, they see it as music and they like it, and it's the same thing with radio, that's if anyone has the vision to see this fact.

Even though Mitch Faulkner's new schedule has kept him very busy with his production company and he has recently left the position as a local disc jockey in the Atlanta area, he feels that production has given him the opportunity to be more creative than format radio has and through many of the seminars which he has given he feels that those interested in becoming the super disc jockey should realize that there aren't many super star positions in radio; but, they should always look at other positions available such as production, programmers, music directors, and technicians.

As Mitch looks back on the black AM disc jockeys of today as opposed to the black AM disc jockeys of the past he recognizes that there isn't much difference in their style but because of their limitations in revenue they haven't had the opportunity to offer some of the things they may have wanted to. And if people would take another look at AM, then, they would find that AM is the music of the youth because the youth get a chance to hear what they want as they have in the past, and as a prime example of their survival AM radio can be found to be popular as we look at those which were established on the local college campuses as they play what the listeners want to hear, and talk about the issues which they are concerned with.

While Mitch Faulkner's passion for the business keeps him going he maintains keeping the bond of fraternal love with his peers in radio and feels that regardless of the color, appearance, etc. of the disc jockey they are (black, white, what have you) filled with motivational love for radio.

While Mitch is always one who's very serious about the industry, he admits that there have been some funny situations which have occurred such as the time when he played Bounce, Kick, Rock 'n' Roll by Vaughn Mason and the crew and as the song began to play he realized that he had to go to the restroom, but, since he had about seven minutes before the song finished, he felt at liberty to do so. Mitch recalls that at WKBD in Kentucky they had the old turn table with the wooden tone arms on, which was real heavy and he didn't know that he had a record with a problem on it, i.e. scratched and at this particular station they didn't have a monitor in the bathroom where he could monitor the broadcast and had not realized that the record was scratch and as he walked back with 1 minute of the song left to play and realized that the song had been scratched for the last

4 and 1/2 minutes as all he could hear was "bounce, bounce, bounce . . . " Immediately, he tried to decide what to do because this was his second job in radio and he was afraid of what the consequences would be, but, he quickly qued up the next song and didn't say anything the rest of his shift. Of course he got a call from the Program Director asking what happened and he didn't know what to do. Mitch said this event took him about three or four years to get over it because it embarrassed him and he wasn't sure how the listeners thought of the bounce overture.

And while radio has been an exciting career for Mitch Faulkner and still is he still remembers one of the great jocks of Atlanta, Scotty Andrews, who wanted him to come back to the station which he had left for another job and before he ever got back to Scotty to let him know his answer Scotty Andrews passed on and he never got a chance to talk with him. When the funeral was held at 1:00 pm Mitch had to be at the station on the air at three, and as he left the church he went to the studio and tried to decide what to do for someone he loved and cared so much for and as he got on the air he decided to give a moment of silence for the chief and had a moment of silence for thirty seconds. As Mitch opened the mike again to the airwaves, the Program Director came in and asked him what happened. Mitch explained and was told that he could have used the thirty seconds for a spot, and as he recalls this was even sadder for him.

And as we close out on Mitch Faulkner's dedication to carry the torch of the pioneers he wants everyone to know that without studying the past of the disc jockey a person won't know where they're going if they don't know where they've been. And as we look at those who are wanting to make it in the industry we must remember to share with each other the history of radio. In order to understand why such a study of the disc jockey is important he relates it to a person who goes out and buys several albums on their favorite entertainer as they get the opportunity to see what they look like, also through the vehicle of videos and concerts they can get the chance to see what they look like, to be in the same room with them, and to get a chance to see them perform, but, with the disc jockey this type of magic moment hasn't been allowed and with the study of the disc jockey this same type of essence can come to pass, along with the parallels of what has happened to some of them in the past which is also happening to some of them now.

And for those of you who never heard of the name Mitch Faulkner, you've heard him all across the United States as he does Voice Overs for the many broadcasters which can be heard from coast to coast across stations all over America.

CONCLUSION

As I begin to close out the story of Black Radio ... Winner Takes All! ... and that of my uncle and his peers it is important that the career plans of those currently involved and those who will choose this path as a future career plan their plight whereas the disc jockey's future is given to those who are willing to live to the standards of those of the past and make the future even more survivable for those to come.

As we examine this pattern of the future wave of radio, it has been found that the best plight would be to study the path of a disc jockey who's career involves the current trend of broadcasting and one who is willing to look at broadcasting from the standard of the past, present, and future.

While we have examined the past of several generations of a people who started out to be called coloured, next negro, and the latter black, and most recently, African American as defined by their skin tones. We can most definitely say that color in most cases became the determining factor used for hiring policies in the beginning, and as time went on, we can also see, that, this was also a measurement for firing, station ownership and selling. The amount of monetary shares received from such a large community of one race, from different regions of the country, has been recorded in the past and the present as we have reviewed the life and styles of some of our earlier pioneers of radio.

Today, we can still see that money is a very important factor in the decision making for hiring and promoting practices, but, this story . . . "Black Radio . . . Winner Takes All! . . . " invited you to see more than the monetary values. `Black Radio . . . Winner Takes All! . . . ' invited you to hear their joy, feel their pain, observe their dedication, and finally to honor their existence. Thank you for letting "Black Radio . . . Winner Takes All! . . . " become a part of your memory and may you share this part of national community history with friends and family as well.

My Voice Became My Cry

My feet became my carriage.
My hands became my guide.
My eyes became my vision.
My voice became my cry.

And though no one could see me
They listened as if they could.
And all the time I cherished
The moments where we stood.
In times of sometimes danger
In times of sometimes war.
I soothed their worn-out spirits.
I soothed them from afar.

It wasn't such a challenge
That I could not control.
The early morning wake-ups,
The passion in my soul.

I gave it all my best shot
I gave it what I could.
And now I've done my duty.
For you, my neighborhood.

Life, now beholds a treasure
Of what we all have shared.
It's not how long I've known you.
It's just because I cared.

Marsha Washington George

PIONEER DISC JOCKEY HONOR ROLL

JOE ADAMS * JOHNNY ALLEN * RON ALLEN * OSCAR ALEXANDER * HERMAN AMIS CARL BOB BAILEY * BOBBY BENNETT * AL BENSON * JIMMY BIRD * LOUISE WILLIAM BISHOP * TEDDY BLACKMON * WILLIAM BLEVINE * DAVE BONDU * MAYME BONDU * JOE BOSTIC * BOB BRISENDINE * DONNY BROOKS * SATELLITE PAPA BROWN * WILLIAM BROWN * RAMON BRUCE BRYANT * SPIDER BURKE * JAY BUTLER * ED CASSLEBURY * CARLTON COLEMAN * ED COOK * JACK COOPER * RALPH COOPER * LUCKY CORDELL * LONNIE CREWS * FRANKIE CROCKER * YVONNE DANIELS * CARL DAVIS * LARRY DEAN * MERRI DEE * HARRISON DILLARD * AL DIXON * DAVE DIXON * OKEE' DO' KEE' * LUCILLE DOUTHIT * DOUG EASON * BEA ELMORE * LEONARD EVANS * JOHN HENRY FORD * MOTHER FRANCIS * KEN GAMBLE * JACK GIBSON * FLASH GORDON * HEATH GORDON * PHIL GORDON * GRACIE * ROCKY GROCHE * FRANKIE HALFACRE * ED HALL * JAMES HAMLIN * FRED HANNA * JOHN HARDY * NIGHT HAWK * BILL HAWKINS * KEN HAWKINS * BERNIE

HAYES * JUANITA HAYES * GORDON HEATH * DOUGLAS HENDERSON * RAY HENDERSON * ERVIN HESTER * ROBIN HOLDEN * JOE HOWARD * MAURICE HULBERT * HAL JACKSON * REV. JOHNNY JACKSON * BOBBY JAY * AL JEFFERSON * WAYNE JOELL * VERE JOHNS * BURKE JOHNSON * IRENE JOHNSON * ROOSEVELT JOHNSON * CASANOVA JONES * E. RODNEY JONES * TOM JOYNER * LONNIE KAY * JIM KELSEY * HERB KENT * MAX KIDD * MICHAEL KIDD * CAROL KING * ADRIAN KENNETH KNIGHT * CHATTY HATTY LEAPER * NIKI LEE * LEON LEWIS * DIZZY LIZZIE * GEORGE LORENZ * JACK LORENZ * C. GENEVA LYMAN * JOHN MARTIN * WILLIE MARTIN * MARY MASON * ANDRE MAURICE * RALPH McIVER * SID McCOY * LARRY McKINLEY * BRUCE MILLER * WILLA MONROE * ANDRE MONTELL * NATHANIEL MONTAGUE * GATE MOUTH MOORE * MANNY MORELAND * CARLTON MOSS * JUGGY MURRAY * JOLLY JOE NORSLEY * DADDY O'DAD * EDDIE O'JAY * ALLEY PAT * EDMUND PATTERSON * ROLAND PORTER * GENE POTTS * BILL POWELL * MARGARET REEVES * REV. I.H. GORDON * KING RO' * BOBBY ROBINSON * EVELYN ROBINSON * JACKIE ROBINSON * WALLY ROCKER * RUDY RUTHERFORD * BUGG SCRUGGS * EARL SELLERS * CURTIS SHAW * DALE SHIELDS * TOMMY SMALLS * MILTON SMITH * PERVIS SPANN * RICHARD STAMPS * DOUG STEELE * SHELLY STEWART * DANNY STILES * HAMP SWAIN * TALL PAUL * CARRIE TERRELL * BERTLEY THOMAS * ROBERT THOMAS * EDDIE 3 WAY * BEA TIBBS * JACK WALKER * JOE WALKER * GEORGE WARD * IRENE WARE * WILEY * WILLIAM (BILL) WILLIAMS * BILL WILLIAMS * BILL (THE WILD CHILD) WILLIAMS * CELENA M. WILLIAMS * JAMES E. WILLIAMS * LARRY WILLIAMS * NAT WILLIAMS * DINO WOODARD * GEORGE WOODS * ED WRIGHT

FAIR PLAY COMMITTEE FACT SHEET

MIAMI SPECIAL
As developed by Disc Jockey Mookie.
DID YOU KNOW:

When the Black jocks at Radio Station WNJR in New Jersey went on strike $3,000.00 was collected to help them out. This money was turned over to officers of NATRA. Each of the seven jocks at WNJR received $50.00; a total of $350.00.

The only people supporting the jocks from WNJR during their strike was personnel from the FAIR PLAY COMMITTEE. Through this COMMITTEE local support was organized in the Black community and a selective buying boycott aimed at advertisers who used WNJR was initiated. This boycott was so effective WNJR sent their executive vice president (from Atlanta, Ga.) to New Jersey to meet with officials of the FAIR PLAY COMMITTEE.

In an effort to destroy the boycott against WNJR's advertisers the station had the FAIR PLAY COMMITTEE brought up on charges before the National Labor Relations Board. After extensive investigation NLRB found the FAIR PLAY COMMITTEE "NOT GUILTY" and the charges were dismissed.

As a direct result of the FAIR PLAY COMMITTEE'S pressure on WNJR the Station Manager was fired.

The first 6 points of our 10 point program has been adopted by station WNJR; including the upgrading of two Black men to executive management positions.

Frank Ward, former manager of Station WWRL in New York City was fired as a result of the FAIR PLAY COMMITTEE'S demand that this be done. In addition two blacks which supported organized gangsters were also fired.

Egmont Sonderling, owner of the Sonderling radio chain, of which WWRL is a member, flew from Los Angeles to New York to meet with FPC officials in an effort to head off a proposed boycott of that stations' advertisers. It was at this meeting that Sonderling reveled he had fired Frank Ward as requested.

Jack Walker of station WLIB in New York was about to be fired three weeks before NATRA convention last year. Due to intervention by members of the FAIR PLAY COMMITTEE he was not only kept on, but given a promotion to become the station's music director.

Radio station WWRL had 6 Black people on their payroll before the FAIR PLAY COMMITTEE stepped in. Five of these people were disc jockeys and the other a janitor. Since that time Black people have been hired on all levels of the stations' operations.

Constant pressure on WWRL paid off when a Black man was named Music Director recently, replacing Larry Berger.

Dee Jay Trivia

1. In 1950 most black disc jockeys were paid the figure of $25.00 salary a week.

2. Frankie Crocker started in radio at the age of about 14 or 15, but, another black disc jockey beat his record by starting at the age of 13 and that was Donny Brooks.

3. Jack Gibson got into radio as a result of not being able to get his career started in acting, because, he was too light to be considered black and not light enough to be considered white, so the industry felt that there was no place for him to fit in, unfortunately, they missed on grasping hold to a legend.

4. In Jacksonville, Florida a street is named Ken Knight Drive after the city's well known community oriented disc jockey.

5. Martha Jean Steinberg, better known as, 'Martha Jean the Queen' once had a release through Epic records entitled, 'Reach Up and Touch Your song'.

6. There were more than three disc jockeys with the name 'Butterball' because the owners owned the names.

7. Two stations that were shut down because of disc jockeys rebutting with management were WJLB in Detroit in 1969 and another in Jacksonville, Florida.

8. Once WVON in Chicago held a contest which announced that $1,450 were hidden in the Washington Park area and the winner could have the money if found, and as the park area was literally destroyed by searchers, the city council held a special meeting and developed an ordinance on their books where no radio station would ever be able to have a contest such as this again.

9. Even though she wasn't a broadcaster, Geneva Lyman was recorded as the first negro female to ever receive a license by the FCC in 1935.

10. The first black disc jockey was Jack Cooper in Chicago, Illinois in 1929.

11. The first colored radio reporter from Harlem was Vere Johns in 1932.

12. There are two disc jockeys named Bill Williams and one other Bill Williams who was an amateur radio operator who used his home base to report to families, the red cross and the media, disaster updates of the worse flood ever seen in the state of Nebraska as recorded in 1935.

13. The first black owned radio station was bought in Atlanta, Georgia in 1949.

14. Hal Jackson has been on the air for more than 50 years and still broadcasts today out of New York.

15. Shelly Stewart out of Birmingham, Alabama has been on the air for more than 50 years and still can be heard today.

16. Doing the 60's era teenagers could call into the station and tell one of their favorite disc jockeys 'good night' . . . many can recall them calling in saying . . . "good night, demon."

17. In the 60's and early 70's when all of the local black am stations signed off, you could easily tune into Hos Allens' station around 10:00 pm and hear all of your favorite tunes. An example of the power of his tower is reaching a distance from Nashville, Tennessee to Jacksonville, Florida.

18. Sometimes disc jockeys would create fun contest for callers, but, these type of contest required your attention, one well known contest was the counting how many time the word daughter was sang in a song called 'daughter, you better leave them boys alone' and when the disc jockey played the song you had to call in with the correct count.

19. In 1958 Alma John's fan club out of New York City brightened the famous Savoy ballroom with 400 members of the Alma John Fan club at their annual wing ding, and her program from radio station WWRL, "What's Right with Teen-Agers," won her the important McCall Award to Women in Radio and Television in 1957 for "service primarily for youth", known as the McCall's Mike.

20. In 1948 George Hudson became the first black staff announcer for New Jersey Radio station WJLK from the Asbury Park area and later moved over to radio station WNJR in Hackensack, New Jersey where he did the morning newscasts for the station until his death in 1970.

21. An article from: Date Line Music City by Charles Lee Lamb reported a contest for the Smothers Brothers song 'Slithery Dee' with a contest called, "What's Slithery Dee?". Winnings were for a four day outing at the Fairmount Hotel in San Francisco. The

contest was open to Dee Jays and their listeners. (Dee Jays and their listeners can like the Smothers Brothers new mercury album, "It Must Have Been Something I said" by making drawings of a Slithery Dee to win which is sponsored by the Smothers Brothers and Mercury Records by contacting Morris Diamond).

Works Consulted

Chicago Historical Society. Jack Cooper. Box 1, Folder 12. Chicago, Illinois.

Henderson, Billie. By This Action Negro Makes Step Forward In New Realm. Pittsburgh, Pa.: The Pittsburgh Courier. 1930.

Radio Sketches by Moss. New York, N.Y., Afro-American. 1931.

Entertainers Boosting Ranks of Baltimore's Own Radio Colony. Pittsburgh, Pa.: The Pittsburgh Courier. 1932.

Vere Johns First Colored Harlem Radio Reporter. Pittsburgh, Pa. The Pittsburgh Courier. 1932.

Joe Bostic—Voice of the Negro Community. New York, N.Y.: Afro-American. 1932.

Daniels, Judy. Radio Artist—John Henry. New York, N.Y.: The New York Age. 1933.

Ripley, Robert. Builder of the Tiniest Set. The Afro-American. 1934.

Radios Warning of Flood. St. Louis, Mo., The St. Louis Argus. 1935.

Radio Operator. The New York Amsterdam News. New York, N.Y., 1935.

Conducts All-Negro Radio Hour. Washington, D.C.: The Washington Tribune. 1936.

Daniel, J.A. . N.Y. Amateur Finds 300 Colored Radio Operators. New York, N.Y.: Afro-American. 1936.

They All Make Good But Cliff. Herald Tribune. 1941.

Morrison, James. Success & Clifford Burdette. Negro Radio Impresario at 27. New York, N.Y.: The Daily Worker. 1941.

McManus, John T. . WOR Refuses OWI Negro Program. New York, N.Y.: Afro-American. 1943.

Blom. Heard and Over Heard—WMCA Scores Again. New York, N.Y.: Daily Worker. 1945.

Peck, Seymour. Corwin, Walter White in 'Crisis in Radio' Confab. New York, N.Y.: The New York Times. 1947.

Wednesday Nights Bring "Swingtime at the Savoy". New York, N.Y.: The New York Times. 1948.

Carson, Saul. Lee, Robeson Blast Bias in Radio. Pittsburgh, Pa.: The Pittsburgh Courier. 1949.

Lauter, Bob. WMCA Proves A Point. New York, N.Y.: The Sunday Worker. 1949.

Robinson, Major. Radio Revolt In The South. Our World. 1949.

Robinson, Major. Blues, Bebop Saved WDIA From Bankruptcy. Negro Voices Influence Sales. Our World. 1949.

Lauter, Bob. F.C.C. Suffers Curious Attack of Sensitivity. New York, N.Y.: The Sunday Worker. 1950.

Winston-Salem Opens Negro Radio Station. Winston-Salem, N.C.: The Amsterdam News. 1950.

Platt, David. Startling Facts About TV, Radio. New York, N.Y.: The Sunday Worker. 1954.

Tuck, Jay Nelson. National Radio Network for Negroes To Open With 40 Stations Jan. 18. New York, N.Y.: New York Post. 1953.

Away From The Blues. New York, N.Y.: Newsweek. 1954.

Black Deejay Quits Station In Fuss Over Record. Jet Magazine. 1968.

Cordell, Lucky. NATRA Files and Documentation received from Past National President. 1960's and 1970's files. Chicago, Ill.

Dallos, Robert E. Black Radio Stations Send Soul and Service to Millions. New York Times. 1968.

Radio/Race Relations Information Center (1970). Nashville, Tenn.: 1970.

Carlson, Doug. Blind Sportscaster 'sees' action for radio listeners. Atlanta, Ga.: Atlanta Constitution. 1986.

The Negro Almanac Reference Work on the African American. 5th Edition. Leon Lewis—Radio Commentator. Gale Research Inc., 1989.

Jones IV, James T. . Pioneer DJ sees the past in radio's future. New York, N.Y.: USA Today. 1989.

Hinckley, David. Hal Jackson and Fifty Years of Black Radio History. New York, N.Y.: Daily News. Sunday Morning Classic. 1990.

Interview Credits

Pioneer Disc Jockey—Les Anderson
Pioneer Disc Jockey—Donny Brooks
Pioneer Disc Jockey—Paul E.X. Brown
Pioneer Disc Jockey—Jay Butler
Pioneer Disc Jockey—Eddie Casslebury
Pioneer Disc Jockey—Charlie Derrick
Pioneer Disc Jockey—Eddie O'Jay
Pioneer Disc Jockey—Jack 'The Rapper' Gibson
Pioneer Disc Jockey—Frank Halfacre
Pioneer Disc Jockey—Casanova Jones
Pioneer Disc Jockey—E. Rodney Jones
Pioneer Disc Jockey—Herb Lance
Pioneer Disc Jockey—Bill Doc Lee
Pioneer Disc Jockey—Alley Pat
Pioneer Disc Jockey—Willis Scruggs
Pioneer Disc Jockey—Dale Shields
Pioneer Disc Jockey—Shelly Stewart
Pioneer Disc Jockey—Richard Stamz
Pioneer Disc Jockey—William Summers
Pioneer Disc Jockey—Joe Walker
Disc Jockey —Bill Baker
Disc Jockey
And Producer —Mitch Faulkner
Disc Jockey —Tommy Smalls, Jr.
Disc Jockey —Doug Steele
Disc Jockey —Bill Williams
Disc Jockey & Instructor—Johnny Allen
Pioneer Record Promoter—Gene Burleson
Pioneer Record Promoter—Bunky Shepard
Pioneer Record Promoter—Joe Medlin
Pioneer Record Promoter-Melvin Moore

INDEX

"Bouncing With Bobby" 72
"Careless Love," 42
"Castle Rock Show" 119
"Chatty Hattie" Leaper 119
"Crown Prince of Disc Jockeys" 198
"Diggy Do" 97
"Dizzie Lizzie" 149
"early bird" disc jockey 64
"Had A Call" 140
"Hi Neighbor" 72
"How Soulful is `Soul' Radio?" 203
"John R" 198
"Life Begins at Midnight" 191
"Lucky's Soul Kitchen" 108
"Mr. 1/2" 107
"Mr. Lucky" 107
"Negro oriented format" 51
"OKY-DOO-KEE" 97
"Shelly The Playboy". 96
"Slaughters Restaurant" 147
"The All-Negro Radio Hour," 55
"The Bob Cat" 166

'All-Negro Radio Hour' 55
'Baby I'm For Real'—By the Originals 139
'Bad Boys' 59
'Burn Baby Burn' 99
'Headlines In Review' 70
'Jack The Rapper-Newsman of the Year Award 124
'My Baby Left Me...Standing In The Back Door Crying... 85
'OK' group 97
'Open The Door Richard' 131, 49
'Quiet Village' 115
'race' records 146
'Say It Loud, I'm Black and I'm Proud' 109
'Shelly The Playboy Fan Club' 96
'shred the boards' 88
'The Crackers' 71
'The Midnight Ramble' 93
'The Sound' 93
'Why Dial Another Station?' 114
'100% Wrong Club' 158
1929 Rose Bowl game 15
a Miss Black America pageant 58
A.D. Walden 71
A-Tisket, A-Tasket 26
ABC 169
Abraham Lincoln 47
Address Unknown 43
advertise 14
Afro-American 226, 227
AFTRA 101, 104, 121
Age Herald 120
air jockeys 19
Al Benson 52, 105, 106, 136, 145, 152, 153, 196, 217
Al Donohue 26
Al Jarvis 16

Al Jefferson 68, 174, 198, 218
Alabama House of Representatives 124
Alabama Jazz Hall of Fame 124
Alexander's Ragtime Band 26
Ali and Frazier 160
Allan Freed 198
Alley Pat 110, 218
amateur radio license 186
Ambrose Smith 45
American Broadcasting Company. 15
Amos 'n' Andy 39
Amsterdam News 120
Anita Baker 58
Ann Southern 51
Annual Jazz Hall of Fame 119
Apollo Theater 21, 57, 192
ARB numbers 15
Arbitron Rating of Men and Women 79
Aretha Franklin 109
Ariba 148
Arts Fellow Building 84
Atlanta Association of Black Journalists 158
Atlanta Black Crackers 71
Atlanta Coca-Cola Bottling Company 72
Atlanta University 83
Atlanta's little Harlem 83
Atlantic Brewery 111
Auburn Ave 111, 83
Ave Maria Hour 47
B.B. King 109
Baby Chick 109
Baily and Natalie Hindesas 122
bankruptcy 63, 64, 71, 227
Ben Bernie 15

Ben Hooks 91
Benny Goodman 26
Berry Gordy 58, 93, 117, 146, 149, 197
Bert Ferguson 64
Bertha Idaho 44
Bessie Smith 85
Bessimer, Alabama 98
Beulah 207
BIAS IN RADIO 227
Big Joe Turner 59
Big Mama Thorton 85
big network shows 15
Bill Baker 79
Bill Board 19
Bill Crane 114
Bill Hawkins 147
Bill Randle 16
Bill Summers 117
Billie Holiday 190
Bing Crosby 26
Bing Crosby and Connee Boswell 26
Bird Street 83
Birdland 193
Birmingham Age Herald 206
Birmingham News 69
black baseball teams 49
Black Broadcasters Hall of Fame 124
Black Music Association 178
Black Radio Hall of Fame 99, 169, 198
Blind Sportscaster 158, 228
Blue Network 14
Blue Velvet 93
Blues Foundation 87
Bo Diddley 109

Bob Brisandine 72
Bob Perkins — Young Widow Brown 88
Bob Roundtree 166
Bobby Blue Bland 59
Bobby Brown 87
Bobby Jay 189, 196, 218
Bobby Massey 149
bombings 16
Boston 20, 186
British West Indies regiment 43
brokerage arrangements 38
Brooke Benton 117
Brooklyn 20, 71, 116, 160
Buck Barnes 44, 46
Buddy Deane 66
Buddy King 100
Bull Moose Jackson 190
Bull of the Woods Chewing Tobacco 92
Butterball 98, 114, 137, 140, 174
Buttermilk Bottom 90
C. Geneva Lyman 188
Cab Calloway Quizzeale 207
Canton 148
Carlton "King" Coleman 194
Carlton Moss 42, 218
Carrie Terrell 218
CBS 15, 47, 49, 56, 122, 129
censorship 20, 21
Chadburn, N.C. 23
Champion Enterprises 169
Change Partners 26
Charles Brown 143
Chattanooga, Tennessee 71
Chicago World 69

Chicago's 'First Lady of Radio' 142
Chuck Berry 108
Chuck Richardson 44, 45
Church of Incarnation 158
Cisco's 176
Clarence Williams 42
Clifton Spriggs 44
coast to coast network broadcast 15
Coasters—Yakety Yak 20
Coca-Cola 72, 191
Cohen brothers 79
colored radio operators 226
Columbia Phonograph Broadcasting Company 15
Columbus, Georgia 98
Commodores 78
Communications Act 22
Community Players in "Romeo and Juliet" broadcast 134
Conneaut, Ohio 108
Conrad Mallock 169
console sets 15
Cooks 83
cotton fields 65
Count Basie 58, 85, 104
Crayton's Sausage firm 147
Curtis Mayfield 58
Danny Stiles 359, 367, 418
Dave and Mayme Bondu 217
Dave Clark 87
Dave Crawford known as 'The Demon' 114
Dave Dixon 114
Dave Galloway 89
David Platt 120
Daytona 74, 77, 80, 89
Decatur 71

Deluxe Radio Theater 88
Dempsey-Carpentier fight 14
Denver, Colorado 186
Depression Era 69
deputy commissioner of sanitation 194
Detroit, Michigan 138
Dick Kemp 154
Dick Saunders 81
Dictograph Fire Safety Products 147
Dinah Washington 85
Dionne Warwick 58
Dixie airways 81
Don Lee Cadillac building 15
Donny Brooks 113, 217
Doug Carlson 158
Doug Steele 175, 176, 218
Douglas "Jocko" Henderson, the Rocket Ship man 192
Dr. Bell 167
Dr. Frank Conrad 13
Dr. Jive 21, 116, 131, 194
Drifters—Save The Last Dance For Me 20
Duke Ellington 89, 104, 134, 146, 168, 192, 206
Dumont 121
Dwight Qawi 159
E. Rodney Jones 100, 101, 102, 103, 104, 105, 110, 113, 218, 229
Eartha Kitt 173
Ed Bradley 114
Ed Cook 137, 198, 217
Ed McKenzie 16
Ed Medlin 71
Ed Reynolds 95
Ed Smalls 116
Ed Sullivan 116
Eddie Castleberry 92, 118, 119, 123, 126

Eddie Ernester 69
Eddie Levert 149
Eddie O'Jay 79,127, 145, 192
Edgewood Avenue 98
Edward R. Murrow 207
Egypt 43
Elton John 140
Elvis 192
Ernie's Record Mart 109
Ethel Waters 207
Ethical Culture-Society School 47
Etta James 59
Etta Jones 117
Europe 46
Eva Taylor 42
Evander Holyfield 159
Evelyn Robinson 191,193, 194, 218
fan mail department 15
Farmers Almanac 94
Fats Domino 197
Fats Waller 26
Federal Communications Commission 22, 182, 203
Feed the Hungry Telethon on WGNX 158
first Colored Harlem Reporter 46, 226
first commercial advertising deal 14
first live orchestra program 14
Florida A&M University 160
Florida School for the Deaf and Blind 15
Fordham University 171
Frank Sinatra 168
Frank Wilson 42
Frankie Crocker 107, 195, 217, 218
Frankie Halfacre 107, 108, 128, 217
Frankie Laine 190

Franklin McCarthy 137
Fred Astaire 26
Fred Woodress 120
G.I. Jive 29
Garden of Time 47
Gatemouth Moore 64
George "Hound Dog" Lorenz 198, 218
George Carter 65
George White 167
Georgia Burke 42
Georgie Woods 114
Gladys Knight 94, 109
Gleason's Musical Bar 148
Globe Trotters 79
Godfrey Harris 45
Goldfinger movie sound track 114
Gordon Heath 46, 217
Graham McNamee 15
great depression 15
Greenville, South Carolina 123, 124
Guy Lombardo 190, 206
H.B. Barnum 149
Hal Jackson 16, 21, 57, 189, 193, 196, 197, 198, 218, 222, 228
Hamilton Music Store 14
Hancock candies 147
Harding-Cox Presidential election 14
Harrison Dillard 147
Hawks games 161
Hearst 122
Here Comes Tomorrow 88
Herman Amis 194
Herndon Building 84
High Five from Chicago 112
High Sumner High School 69

Hill Street 82
hit records 16
Hollywood 14, 88
Hooper survey 64
Hoover vacuum cleaners 191
Horse Jockeys 92
Hot Rod 65, 66, 67, 68, 69, 194
Hot Rod Hulbert 64
Hotel Theresa 193
Hoyt Sullivan Hair Products 109
Hunter street 94
I Let a Song go out of my heart 26
Ike and Tina Turner 22
Ike Dixon 45
Ink Spots 85
Inner City Broadcasting 58
Inspirations Across America! 177
Irgles 92 Beer 92
Irvin Hughes 45
Italian Swiss Colony Wine 148
Italy 138
J.B. Blayton 71
J.L. Doss 96
Jack Cooper 36, 37, 38, 39, 40, 41, 127, 217, 222, 226
Jack Springer 114
Jack The Rapper 48, 49, 50, 51, 54, 86, 87, 88, 89, 98, 124, 125, 126, 128, 177, 178, 179, 208
Jack Walker 193, 194
Jackie Robinson 218
Jackie Wilson 90, 116, 189
Jackson 5 58
James Brown 93, 109, 134, 204
James Coles 44
James E. Williams 218

Jay Butler 138, 139, 140, 141, 217, 229
Jerome Washington 44
Jesse Jackson 91, 124
Jessie Owens 'the Olympic Runner' 153
Jet Magazine 90, 228
Jim Crow 120
Jimmy Bishop 114
Jimmy Lunsford 89
Jimmy Reed 109
Jock O' Henderson 192
Jockey Jack 22, 86, 89, 93, 128
jockey suit 92
Joe Bailey 80
Joe Bostic 44, 46, 47, 194, 217, 226
Joe Howard 112, 166, 167, 198, 218
Joe Lewis 91
Joe Medlin 87, 229
Joe Walker 128, 157, 158, 159, 218, 229
Joe Williams 85
John Charles Thomas 14
John Henry Ford 217
John Lee Hooker 111
John Martin 218
John Pepper 64
John Richbough 109
John Sterling 161
John Wesley Dobbs 70, 71
Johnny Allen 172
Johnny Carson's tower 115
Johnny Christian 45
Johnny Patterson 107
Johnny Shaw 76, 80, 114
Jolly Joe Norsley 139
Joltin' Joe Howard 198

Joseph Horne Co. of Pittsburg 14
Joyce Jenkins 88
Juanita Hayes 147
Jumpin' Joe Howard 112, 166, 167, 198, 218
KAAY 100
Karamu House 147
KATZ 181
KCBS 14
KDKA 14
Ken Knight 9, 16, 49, 72, 73, 74, 75, 76, 77, 78, 79, 80, 81, 89, 91, 95, 98, 110, 111, 116, 128, 129, 156, 161, 146, 154, 158, 187, 188-190, 197, 252, 262, 338, 419
Ken Knight Drive 222
KFWB 16
King Pleasure's, "Moods' Mood for Love" 107
Korean War 107
KYOK 115
Larry Dean 117
Larry Dixon 168
Larry McKinley 198
Larry Picus—'Jack The Bell Boy' 114
Larry Williams 218
late Senator Bilbo 82
Laurie Gomez, the All American Distant Runner of North Carolina University 108
Lemonyes Owens College 50
Leon Lewis 116, 169, 170, 171, 172, 218
Leon Nelson 44
Leonard Chess 135, 154
Leonard Evans 172
Life is Beautiful 88
Lincoln Country Club 110
Lithonia 93
Lloyd Fonteroy 114

London 16, 138
Lonely Tear Drops by Jackie Wilson 90
Los Angeles 116, 138, 143, 144, 189
Louis Jordan 190
Louise Williams Bishop 172
Louisville, Kentucky 92
Love Walked In 26
Lucius Barnett 108
Lucky Cordell 106, 108, 131, 151
M&M 176
Madison Square Garden 58
Magnificent Montaque 99, 137
Magnolia on Sunset 70
mail count 15
Majestic Hotel 147
Major Robinson 81
Malcolm Moore 98
Manny Moreland, Jr. 145
Martha Jean Steinberg 98, 99, 114, 140, 222
Martin Block 16
Martin County, Fla. 164
Martin Luther King 21, 91, 135, 162, 198
Mary Dee 130
Mary Wells 90, 126
Mascots 148
Maynard Jackson 70
Mayor of Harlem 171, 193
MCA 169
Memphis, Tennessee 50, 99, 135
Miami, Florida 99, 124
Michael Jackson 57
Michigan State 49
Midway Radio & Television School 83
Milton "Butterball" Smith 198, 218

Milton Cross 14
Moody Bible Institute 69
Morgan College 42
Motown 58, 87, 90, 93, 102, 117, 146, 149, 177, 197, 210
Mozart 95
Mr. Hardin 71
Mr. Shipman 116
Muddy Waters 111
Music, Maestro, Please 26
Mutual Black Network News 124, 125
N.Y. Amsterdam News 171, 226
NAACP 53, 61, 119, 123, 156, 158, 178, 185, 208
NARA 112, 179, 202
Nat King Cole 148, 193
Nat Williams 63, 64, 218
National Broadcasting Company 14, 60, 179, 219, 220, 228
NATRA 50, 101, 104, 112, 114, 117, 156, 179, 219, 220, 228
NBC 14, 15, 61, 122
Negro CPA 83
Negro Hour 38
Nelson Mandela 87
Network radio 14
New Edition 58
new jack swing 58
New Orleans, Louisiana 99, 100, 101, 135, 160, 198
New Theater School 47
New World A-Coming 46
Nipsey Russel 85
Norman Spaulding 39
NYA (National Youth Administration) projects 47
Olympics 60
On Air Performer 50
Open Mike Entertainment magazine 77
Orchestras 45

Oscar Alexander 198
Otis Redding 109
Our World 82
Our World Publication 81
Owen Dodson 47
Paradise West 116
Pat Boone 198
Patti Page 192
Paul E.X. Brown 69, 70, 71, 72, 96, 131, 229
Paul Williams 190
Pearl Primus show at the Belasco 190
people of color 37, 42, 44, 172
Pepsi Cola 152
Percy Glascoe 45
Percy Sutton's Inner City Broadcasting Corp. 189
Personality Jocks 91, 192
Phil Gordon 194, 217
Philip Morris 206
Pied Pipers of Rock 'n' Roll 198
Pittsburgh Courier 72, 226, 227
Playboy Records 138
popular records 16
Post-War Disc Jockeys 41
Postal Carriers 83
pre-quiet storms 93
President Franklin Roosevelt 16
Public Affairs 22, 24, 67, 72, 171, 203, 205
Pullman cars 83
Pullman Porters 83
Pulse 15, 120
Queens 116
Radio Corporation of America 14
radio correspondents 16
radio drama 88, 151

radio receiving sets 14
radio telephone 14
RADIOS WARNING OF FLOOD 223, 226
Rameses Rhythm Boys 44
Ramon Bruce 194, 202
Ray Charles 143, 160, 168
record librarian 70
record promotion 14, 100
Red Cross Chief 223
Regal Theatre 22, 102
Reuben Parker 44
Reverend I.H. Gordon 147
Richard Stamz 49, 50, 51, 53, 54, 56, 131, 132
Ringling Brothers Circus 79
Rivers Chambers 45
RKO movie lot 15
rock 'n' roll 17
Rock and Roll Hall of Fame 198
Rocky G. George Hudson 194
Roland Porter 137, 139, 218
Rome 138
Rose Bowl 15
Rounsaville Broadcasting 23, 123, 204
Roy Wood 47, 48, 104, 154, 109
Royal Crown Hair Dressing 109
Rudy "The Deuce" Rutherford 198
Rudy Valee 15
Ruth Brown 111
Salesman 50, 148,171
Sam and Dave 150
Savoy Ballroom 207, 224, 227
Schomburg Collection of the New York Public Library 46
SCLC 158
Selena M. Williams 147

Shelly Stewart 71, 95, 99, 218, 223, 229
Shorty Moore 145
Showtime at the Apollo 58
Sid McCoy 52, 141, 142, 218
Smithsonian Institute 95, 124, 125
Soft Sheen 176
Soul Coast to Coast 175
Soul Finger 114
South Africa 87
southern radio revolt 82
Speidel Broadcasters, Inc. 23
St. Louis Argus 226
Stanley Ray 97
Star Broadcasting Company 115
Startling Facts About Radio in 1954 120
State Penitentiary 95
station 8XK 14
Stax 87, 169
Stax Records 87, 169
Steiner Brew 111
Stevie Wonder 57, 58, 87, 93, 102, 134
Sugar Lump 17
summer theater 88
Sunday Morning Classics 189
Supremes 93, 197, 210
Sweet Auburn 84
Sweet Pea Snuff 92
Sybil 58
Terry Taylor 108
the 'hitmaker' 109, 115
The AFRO 226, 227
The Age 43
the Big Daddy of R&B Disc Jockeys 109
the blues 189, 193, 206

The Call & Post Ball Room 148
The Chicago Giants 51
the Cleveland Browns 108
the construction permit 115
The Dells 117
The Detroit News 14
The Florence Ballard Award 92
The Leaner Brothers United Distributors 146
The National Association of Broadcasters 198
The OK chain 97
the Olympic Gold Winner 147
the Orioles 66, 190
the Palm 193
the Ravens 190
the Royal Peacock 84
The Shadow 88
The Southernaires 42
the Spaniels 189
The Temptations 57, 67, 68, 78
the Today Show 89
The Top Hat 84
The Twenty Harlem Fingers 45
Tom Delaney 45
Tom Duncan 152
Tommy Cowan 14
Tommy Smalls 16, 115, 116, 117, 131, 189, 218
Top 40 18, 142, 189
trade out 14
Trudy Haynes 168
Trust Company 71
TV sets 15, 121
Ulysses Chambers 46
United Artists 138
United Independent Broadcasters 15

University of Minnesota, Columbia College 69
urban ghettos of the North 41
Valley Forge Army Hospital 107
Variety 192
Vee Jay 79, 146
VEGA 117
Vere Johns 43, 218, 223, 226
Victorola 14
Vincent Lopes 14
Virgin-Islands 138
Vivian and Jimmy Brackens 146
WAAT 121
WABC 122
WABQ 123
WADO 192
Walt Whitman 47
Walter Eglin Jr. 96
Walter Williams 149
WAOK 72, 97, 98, 110, 118, 119, 158, 161
WASH -FM 124
Washington Tribune 226
WATL 72
WATV 98
WBBR 113
WBCO 98
WBLK 115
WBLS 58, 189, 195
WCAN 145
WCAO 44
WCBM 44
WCBS 15
WCBW 47
WCFL 141, 142
WCHB 23, 166, 167, 204

WCIN 19, 123, 124, 126
WDAS 114, 202
WDCA 43
WDIA 63, 64, 65, 180, 198, 227
WEAF 14, 42
WEAS 71, 72, 75
WEBB 23, 44, 68, 124, 204
WEBR 44, 45
WEDR 71, 95, 96, 98, 99, 120, 123, 124
WENN 48, 98
WERD 70, 72, 74, 77, 79, 80, 82, 83, 84, 85, 86, 87, 94, 99, 110 131, 158, 162, 163
Wes Smith 198
Westinghouse 14
WEUP 23, 204
WEVD 191
WFIZ-FM 108
WGCI AM 134, 142
WGES 50, 51, 52, 152, 154, 181
WGN 20
WGOK 123, 179
WGRY 134, 153
WGST 161
WGUN 71
WGV 43
WHBI 191
White Rose Petroleum Jelly 109
Whitfield Records 138
Whitney Houston 87
Whitney Young 91
WHOM 122, 193
WIGO 79, 163
Wil Marion Cook 46
Willa Monroe 64

William (Bill) Killibrew 146
William Blevine 55, 217
William Brown 198
William Powell 149
William S. Paley 15
Williams Brothers 176
Willie Bryant 193
Willie Martin 80, 218
Willis Scruggs 84, 85, 229
Wilson Pickett 109
WINS 17, 122, 193, 197
Winston-Salem 198, 227
WINX 58
WIZ 14
WJAK 139
WJJD 87, 134
WJLB 138, 140, 166, 167, 222
WJLD 96
WJOV 69
WJZ 15, 43, 124
WLAC 109, 208
WLDB 174
WLIB 25, 46, 121, 170, 171, 189, 192, 193, 194, 206, 220
WLIX-TV 121
WLOK 97, 115, 135
WLTH 153
WMBM 123, 124, 157
WMCA 46, 47, 121, 170, 227
WMCM 121
WMID 174
WNEW 16, 121
WNJR 192, 194, 219, 220, 224, 229
WNTA 194
WNUA 142

WNYC 47, 61, 62
WOKS 98
WOKY 145
WOOK 174
WOR 23, 204, 227
WORD 79
World War 16, 19, 40, 43, 70, 134
WOV 191, 192, 193, 194
WPDQ 80
WPNX 194
WQBH 1400 AM 138
WQOK 123, 124
WQXR 121
WRNG 158
WSB 94
WSBC 38, 181
WSDM 142
WSGN 56
WSID 123, 124
WSRS 147
WTKL 100
WUST 174
WVEE 159, 161
WVOK 97
WVON 101, 105, 134, 142, 154, 155, 181, 223
WWIN 66, 67, 174
WWJ 14
WWOW 108
WWRL 66, 115, 116, 189, 192, 220, 221, 224
WYLD-AM 100
WYNR 142, 154
Yates 71
YMCA 54, 179

You Must Have Been A Beautiful Baby 26
You'll never Walk Alone by Roy Hamilton 90
Yvonne Daniels 114, 141, 142, 217

BVG